IDENTIFICATION AND CATEGORIZATION OF SABOTAGE TARGETS, AND IDENTIFICATION OF VITAL AREAS AT NUCLEAR FACILITIES

The following States are Members of the International Atomic Energy Agency:

AFGHANISTAN
ALBANIA
ALGERIA
ANGOLA
ANTIGUA AND BARBUDA
ARGENTINA
ARMENIA
AUSTRALIA
AUSTRIA
AZERBAIJAN
BAHAMAS
BAHRAIN
BANGLADESH
BARBADOS
BELARUS
BELGIUM
BELIZE
BENIN
BOLIVIA, PLURINATIONAL STATE OF
BOSNIA AND HERZEGOVINA
BOTSWANA
BRAZIL
BRUNEI DARUSSALAM
BULGARIA
BURKINA FASO
BURUNDI
CABO VERDE
CAMBODIA
CAMEROON
CANADA
CENTRAL AFRICAN REPUBLIC
CHAD
CHILE
CHINA
COLOMBIA
COMOROS
CONGO
COOK ISLANDS
COSTA RICA
CÔTE D'IVOIRE
CROATIA
CUBA
CYPRUS
CZECH REPUBLIC
DEMOCRATIC REPUBLIC OF THE CONGO
DENMARK
DJIBOUTI
DOMINICA
DOMINICAN REPUBLIC
ECUADOR
EGYPT
EL SALVADOR
ERITREA
ESTONIA
ESWATINI
ETHIOPIA
FIJI
FINLAND
FRANCE
GABON
GAMBIA
GEORGIA
GERMANY
GHANA
GREECE
GRENADA
GUATEMALA
GUINEA
GUYANA
HAITI
HOLY SEE
HONDURAS
HUNGARY
ICELAND
INDIA
INDONESIA
IRAN, ISLAMIC REPUBLIC OF
IRAQ
IRELAND
ISRAEL
ITALY
JAMAICA
JAPAN
JORDAN
KAZAKHSTAN
KENYA
KOREA, REPUBLIC OF
KUWAIT
KYRGYZSTAN
LAO PEOPLE'S DEMOCRATIC REPUBLIC
LATVIA
LEBANON
LESOTHO
LIBERIA
LIBYA
LIECHTENSTEIN
LITHUANIA
LUXEMBOURG
MADAGASCAR
MALAWI
MALAYSIA
MALI
MALTA
MARSHALL ISLANDS
MAURITANIA
MAURITIUS
MEXICO
MONACO
MONGOLIA
MONTENEGRO
MOROCCO
MOZAMBIQUE
MYANMAR
NAMIBIA
NEPAL
NETHERLANDS, KINGDOM OF THE
NEW ZEALAND
NICARAGUA
NIGER
NIGERIA
NORTH MACEDONIA
NORWAY
OMAN
PAKISTAN
PALAU
PANAMA
PAPUA NEW GUINEA
PARAGUAY
PERU
PHILIPPINES
POLAND
PORTUGAL
QATAR
REPUBLIC OF MOLDOVA
ROMANIA
RUSSIAN FEDERATION
RWANDA
SAINT KITTS AND NEVIS
SAINT LUCIA
SAINT VINCENT AND THE GRENADINES
SAMOA
SAN MARINO
SAUDI ARABIA
SENEGAL
SERBIA
SEYCHELLES
SIERRA LEONE
SINGAPORE
SLOVAKIA
SLOVENIA
SOUTH AFRICA
SPAIN
SRI LANKA
SUDAN
SWEDEN
SWITZERLAND
SYRIAN ARAB REPUBLIC
TAJIKISTAN
THAILAND
TOGO
TONGA
TRINIDAD AND TOBAGO
TUNISIA
TÜRKİYE
TURKMENISTAN
UGANDA
UKRAINE
UNITED ARAB EMIRATES
UNITED KINGDOM OF GREAT BRITAIN AND NORTHERN IRELAND
UNITED REPUBLIC OF TANZANIA
UNITED STATES OF AMERICA
URUGUAY
UZBEKISTAN
VANUATU
VENEZUELA, BOLIVARIAN REPUBLIC OF
VIET NAM
YEMEN
ZAMBIA
ZIMBABWE

The Agency's Statute was approved on 23 October 1956 by the Conference on the Statute of the IAEA held at United Nations Headquarters, New York; it entered into force on 29 July 1957. The Headquarters of the Agency are situated in Vienna. Its principal objective is "to accelerate and enlarge the contribution of atomic energy to peace, health and prosperity throughout the world".

IAEA NUCLEAR SECURITY SERIES No. 48-T

IDENTIFICATION AND CATEGORIZATION OF SABOTAGE TARGETS, AND IDENTIFICATION OF VITAL AREAS AT NUCLEAR FACILITIES

TECHNICAL GUIDANCE

INTERNATIONAL ATOMIC ENERGY AGENCY
VIENNA, 2024

COPYRIGHT NOTICE

Publishing Section
International Atomic Energy Agency
Vienna International Centre
PO Box 100
1400 Vienna, Austria
tel.: +43 1 2600 22529 or 22530
email: sales.publications@iaea.org
www.iaea.org/publications

Printed by the IAEA in Austria
November 2024
STI/PUB/2092
https://doi.org/10.61092/iaea.74e6-e2yc

IAEA Library Cataloguing in Publication Data

Names: International Atomic Energy Agency.
Title: Identification and categorization of sabotage targets, and identification of vital areas at nuclear facilities / International Atomic Energy Agency.
Description: Vienna : International Atomic Energy Agency, 2024. | Series: IAEA nuclear security series, ISSN 1816–9317 ; no. 48-T | Includes bibliographical references.
Identifiers: IAEAL 24-01716 | ISBN ISBN 978-92-0-120924-5 (paperback : alk. paper) | ISBN 978-92-0-121024-1 (pdf) | ISBN 978-92-0-121124-8 (epub)
Subjects: LCSH: Nuclear facilities — Security measures. | Nuclear facilities — Safety measures. | Sabotage.
Classification: UDC 621.039.58 | STI/PUB/2092

FOREWORD

by Rafael Mariano Grossi
Director General

The IAEA Nuclear Security Series provides international consensus guidance on all aspects of nuclear security to support States as they work to fulfil their responsibility for nuclear security. The IAEA establishes and maintains this guidance as part of its central role in providing nuclear security related international support and coordination.

The IAEA Nuclear Security Series was launched in 2006 and is continuously updated by the IAEA in cooperation with experts from Member States. As Director General, I am committed to ensuring that the IAEA maintains and improves upon this integrated, comprehensive and consistent set of up to date, user friendly and fit for purpose security guidance publications of high quality. The proper application of this guidance in the use of nuclear science and technology should offer a high level of nuclear security and provide the confidence necessary to allow for the ongoing use of nuclear technology for the benefit of all.

Nuclear security is a national responsibility. The IAEA Nuclear Security Series complements international legal instruments on nuclear security and serves as a global reference to help parties meet their obligations. While the security guidance is not legally binding on Member States, it is widely applied. It has become an indispensable reference point and a common denominator for the vast majority of Member States that have adopted this guidance for use in national regulations to enhance nuclear security in nuclear power generation, research reactors and fuel cycle facilities as well as in nuclear applications in medicine, industry, agriculture and research.

The guidance provided in the IAEA Nuclear Security Series is based on the practical experience of its Member States and produced through international consensus. The involvement of the members of the Nuclear Security Guidance Committee and others is particularly important, and I am grateful to all those who contribute their knowledge and expertise to this endeavour.

The IAEA also uses the guidance in the IAEA Nuclear Security Series when it assists Member States through its review missions and advisory services. This helps Member States in the application of this guidance and enables valuable experience and insight to be shared. Feedback from these missions and services, and lessons identified from events and experience in the use and application of security guidance, are taken into account during their periodic revision.

I believe the guidance provided in the IAEA Nuclear Security Series and its application make an invaluable contribution to ensuring a high level of nuclear security in the use of nuclear technology. I encourage all Member States to promote and apply this guidance, and to work with the IAEA to uphold its quality now and in the future.

EDITORIAL NOTE

Guidance issued in the IAEA Nuclear Security Series is not binding on States, but States may use the guidance to assist them in meeting their obligations under international legal instruments and in discharging their responsibility for nuclear security within the State.

Guidance expressed as 'should' statements is intended to present international good practices and to indicate an international consensus that it is necessary for States to take the measures recommended or equivalent alternative measures.

Security related terms are to be understood as defined in the publication in which they appear, or in the higher level guidance that the publication supports. Otherwise, words are used with their commonly understood meanings.

An appendix is considered to form an integral part of the publication. Material in an appendix has the same status as the body text. Annexes are used to provide practical examples or additional information or explanation. Annexes are not integral parts of the main text.

Although great care has been taken to maintain the accuracy of information contained in this publication, neither the IAEA nor its Member States assume any responsibility for consequences which may arise from its use.

The use of particular designations of countries or territories does not imply any judgement by the publisher, the IAEA, as to the legal status of such countries or territories, of their authorities and institutions or of the delimitation of their boundaries.

The mention of names of specific companies or products (whether or not indicated as registered) does not imply any intention to infringe proprietary rights, nor should it be construed as an endorsement or recommendation on the part of the IAEA.

CONTENTS

1. INTRODUCTION

BACKGROUND

1.1. The Convention on the Physical Protection of Nuclear Material [1] provides a framework for ensuring the physical protection of nuclear material, with the 2005 Amendment to the Convention [2] defining sabotage as follows:

> "[A]ny deliberate act directed against a nuclear facility or nuclear material in use, storage or transport which could directly or indirectly endanger the health and safety of personnel, the public or the environment by exposure to radiation or release of radioactive substances."

The Amendment places the obligation on State Parties to the Convention to establish an appropriate physical protection regime applicable to the nuclear material and nuclear facilities under their jurisdiction, with the aim of protecting against sabotage and/or preventing, mitigating or minimizing the radiological consequences of sabotage.

1.2. IAEA Nuclear Security Series No. 13, Nuclear Security Recommendations on Physical Protection of Nuclear Material and Nuclear Facilities (INFCIRC/225/Revision 5) [3], provides recommendations on protection against sabotage while applying the principle of a graded approach on the basis of an analysis of the potential consequences.

1.3. This publication provides guidance on how to implement the recommended requirements set out in Ref. [3] related to sabotage target identification and identification of vital areas. It also supports guidance provided in IAEA Nuclear Security Series No. 27-G, Physical Protection of Nuclear Material and Nuclear Facilities (Implementation of INFCIRC/225/Revision 5) [4].

1.4. This publication builds on the threat assessment presented in IAEA Nuclear Security Series No. 10-G (Rev. 1), National Nuclear Security Threat Assessment, Design Basis Threats and Representative Threat Statements [5], and can be used to provide input for the design of nuclear security systems and measures.

1.5. This publication supersedes IAEA Nuclear Security Series Nos 4, Engineering Safety Aspects of the Protection of Nuclear Power Plants Against Sabotage[1], and 16, Identification of Vital Areas at Nuclear Facilities[2].

OBJECTIVE

1.6. The objective of this publication is to provide technical guidance to States and/or operators for analysing the potential effects of sabotage attacks on a nuclear facility and for implementing actions that can mitigate these effects in accordance with the country's design basis threat or representative threat statement. It also provides guidance on identifying items that need to be protected against sabotage — structures, systems and components (SSCs) at nuclear facilities, associated operator actions[3], and nuclear and other radioactive material — and guidance on the proper operation of these SSCs.

1.7. This publication provides detailed guidance on (a) the identification of potential sabotage targets in a nuclear facility and possible vulnerabilities that could lead to unacceptable or high radiological consequences (see paras 3.3–3.5) if an initiating event of malicious origin were to take place; and (b) the identification of vital areas in nuclear facilities. It also includes guidance to assist States in accounting for the potential risks to a facility associated with stand-off sabotage attacks.

1.8. This publication is intended to be used by States, competent authorities (including the regulatory body) involved in protection against the sabotage of nuclear and other radioactive material, and relevant technical and scientific support organizations, as well as the operators of associated facilities and activities.

---

[1] INTERNATIONAL ATOMIC ENERGY AGENCY, Engineering Safety Aspects of the Protection of Nuclear Power Plants Against Sabotage, IAEA Nuclear Security Series No. 4, IAEA, Vienna (2007).

[2] INTERNATIONAL ATOMIC ENERGY AGENCY, Identification of Vital Areas at Nuclear Facilities, IAEA Nuclear Security Series No. 16, IAEA, Vienna (2012).

[3] Associated operator actions are actions associated with the SSCs performed by the operating personnel.

SCOPE

1.9. This technical guidance is applicable to all nuclear facilities, in particular for cases in which a successful act of sabotage at the nuclear facility could result in unacceptable radiological consequences or high radiological consequences, as defined by the State.

1.10. The process and methodology presented in this publication for sabotage target identification may also be applicable to other facilities associated with radioactive material or high value assets, as well as to SSCs and associated operator actions linked to facility operation and processing of nuclear material.

1.11. Specific physical protection systems and measures are not addressed in detail in this publication. Further information on designing, implementing and sustaining a physical protection system can be found in IAEA Nuclear Security Series No. 40-T, Handbook on the Design of Physical Protection Systems for Nuclear Material and Nuclear Facilities [6].

1.12. The identification of targets for unauthorized removal of nuclear material is not addressed in this publication, nor is response planning to mitigate the consequences resulting from sabotage. These are covered in IAEA Nuclear Security Series No. 14, Nuclear Security Recommendations on Radioactive Material and Associated Facilities [7], and Refs [3, 4].

1.13. This publication refers to cyber-attacks only in relation to the capability of an adversary to use a cyber-attack in support of a physical attack in order to degrade nuclear security measures. IAEA Nuclear Security Series No. 17-T (Rev. 1), Computer Security Techniques for Nuclear Facilities [8], provides detailed guidance on the computer security aspects of protection against cyber-attack.

1.14. The identification of cyber-attack targets that might be affected by sabotage through the maloperation of functions performed by computer based systems important to or related to nuclear safety and nuclear security is outside the scope of this publication. Identification and characterization of computer based systems that might be targeted for exploitation through a cyber-attack that can compromise safety, security or nuclear material accounting and control functions is addressed in Ref. [8]. Reference [8] also provides detailed guidance on the design and implementation of computer security policies, programmes and measures to ensure the protection of computer based systems against compromise.

STRUCTURE

1.15. Section 2 of this publication provides a general overview of the methodology for the protection against sabotage and includes threat statements as input. Section 3 sets out the input that is necessary for the completion of the sabotage target identification process. Section 4 describes the process for developing sabotage attack scenarios. Section 5 outlines the steps to be taken to identify potential sabotage targets and identify candidate vital area sets. Section 6 describes the process of identifying sabotage targets to be protected in the vital areas, and Section 7 provides information on off-site sabotage attacks. Section 8 includes a brief overview of the protection of documentation and information related to the identification of vital areas and sabotage targets. The annexes provide practical examples for the sabotage logic model, facility walkdowns and extreme environmental load evaluations.

2. GENERAL OVERVIEW OF THE PROTECTION OF NUCLEAR FACILITIES AGAINST SABOTAGE

2.1. IAEA Safety Standards Series No. SF-1, Fundamental Safety Principles [9], states that "**The fundamental safety objective is to protect people and the environment from harmful effects of ionizing radiation**". Paragraph 2.1 of IAEA Nuclear Security Series No. 20, Objective and Essential Elements of a State's Nuclear Security Regime [10], states that "The objective of a State's *nuclear security regime* is to protect persons, property, society, and the environment from harmful consequences of a *nuclear security event*."

2.2. Sabotage can have severe consequences, and the success of an act of sabotage could result in conditions that challenge safety objectives and security objectives. The success of an act of sabotage entails the successful defeat of not only the physical protection system, but also the facility safety systems and the accident management provisions. Measures to protect against sabotage should therefore be designed and evaluated in cooperation with nuclear safety experts, nuclear security experts and accident management experts, with the involvement of experts in sabotage protection from various competent authorities.

2.3. Paragraphs 2.4–2.25 provide a brief outline of the process for performing a threat assessment, applying a graded approach for protection against sabotage and identifying sabotage targets and vital areas.

THREAT ASSESSMENT FOR SABOTAGE

2.4. The first step in protecting a nuclear facility against sabotage is to perform a national nuclear security threat assessment, which is an evaluation of existing nuclear security related threats. The assessment process makes use of global, regional and national sources of information to determine the attributes and characteristics of potential adversaries. Detailed information on performing a nuclear security threat assessment is provided in Ref. [5].

2.5. If a performance based regulatory approach is used in the State, sabotage attack scenarios should be developed by the facility operators in cooperation with the regulatory body on the basis of the design basis threat applicable to the facility. Such scenarios should be used to design the physical protection system. If a prescriptive regulatory approach is used in the State, the representative threat statement should be used by the regulatory body to develop sabotage attack scenarios and then to establish the corresponding nuclear security requirements. The goal when establishing sabotage attack scenarios should be to determine a set of credible sabotage attack scenarios that a facility might encounter and to inform those who are responsible for the design of physical protection systems of what measures may need to be considered to meet the nuclear security objectives established in the State's nuclear security regime.

2.6. Sabotage attack scenarios developed on the basis of a design basis threat or a representative threat statement can be categorized as type 1 sabotage threat scenarios or type 2 sabotage threat scenarios as follows:

(a) Type 1 sabotage threat scenarios involve criminal or other intentional unauthorized acts committed by insiders or external adversaries intruding into the facility (with or without insider assistance). Cyber threats blended with physical attacks are included in this category.
(b) Type 2 sabotage threat scenarios involve attacks launched from outside the facility boundary, which do not necessitate the presence of adversaries on-site and have the potential to result in unacceptable radiological consequences or high radiological consequences. Examples of such attacks may include stand-off attacks, such as airborne attacks using drones, missiles or aircraft, as well as attacks using directed energy weapons; blasts from chemical explosions; or the destruction of a heat sink from a distance. Some type 2 sabotage threat scenarios may be addressed by strengthening the facility's physical protection system; however, others can present a significant challenge even if the facility's physical protection system has been strengthened. Type 2 sabotage threat scenarios may include cyber-attacks.

2.7. Sabotage attack scenarios provide a basis for the design of a facility's physical protection system against sabotage. The physical protection system should be capable of deterring, detecting, delaying and responding to criminal or other intentional unauthorized acts as defined in credible sabotage attack scenarios, which are based on the design basis threat or the representative threat statement. Examples of deterrence, detection, delay and response measures can be found in Refs [4, 6] and in IAEA Nuclear Security Series No. 39-T, Developing a Nuclear Security Contingency Plan for Nuclear Facilities [11]. To design an appropriate response to an act of sabotage, immediate operator actions related to the restoration of disabled systems should be taken into account.

2.8. The distinction between sabotage threat scenarios involving the physical presence of an adversary on the site and scenarios not involving a physical presence on the site is made to reflect the differences in the ways in which engineering safety measures are considered to counter each type of sabotage threat scenario. In the case of type 1 sabotage threat scenarios, physical protection measures should be used to prevent adversary penetration, and safety measures should be considered as protection to mitigate the consequences of the act of sabotage. In the case of type 2 sabotage threat scenarios, engineering measures should be used to prevent and mitigate the radiological consequences of an act of sabotage, since physical protection functions (i.e. deterrence, detection, delay and response measures) may have more limited capabilities in this regard.

2.9. The risks associated with sabotage threats that are capable of resulting in unacceptable radiological consequences can be reduced, but they cannot be eliminated. It is unlikely that a single part of the operating organization can implement a system to adequately reduce the risk of sabotage.

2.10. Protection against sabotage attacks can be achieved as a combination of the following:

(a) Defeat of a sabotage attack by the physical protection system, which is designed using scenarios that could lead to unacceptable radiological consequences and were developed on the basis of the design basis threat or the representative threat statement.
(b) Prevention of unacceptable radiological consequences by deployment of safety systems and by accident management.
(c) Effective mitigation of radiological consequences through the implementation of on-site and off-site emergency response actions in accordance with the requirements established in IAEA Safety Standards Series No. GSR Part 7, Preparedness and Response for a Nuclear or Radiological Emergency [12]. In

order to design on-site and off-site emergency arrangements to enable effective mitigation of the consequences of an emergency triggered by a nuclear security event, the results of the threat assessment for sabotage should inform the hazard assessment for emergency preparedness and response purposes (see also Requirement 4 of GSR Part 7 [12]). Further recommendations are provided in IAEA Safety Standards Series No. GS-G-2.1, Arrangements for Preparedness for a Nuclear or Radiological Emergency [13], and further guidance can be found in Ref. [14].

GRADED APPROACH FOR PROTECTION AGAINST SABOTAGE

2.11. Reference [3] states:

> "**Physical protection requirements should be based on a *graded approach*, taking into account the current evaluation of the threat, the relative attractiveness, the nature of the *nuclear material* and potential consequences associated with the *unauthorized removal* of *nuclear material* and with the *sabotage* against *nuclear material* or *nuclear facilities*.**"

Specific recommended requirements for protection against sabotage through a graded approach can be found in section 5 of Ref. [3].

2.12. A State should consider the full range of radiological consequences that could be associated with all its nuclear facilities and should categorize them appropriately, in particular with regard to potential radiological consequences that exceed the limits established by the State for unacceptable radiological consequences (see paras 3.3–3.5).

2.13. Categorization of potential radiological consequences resulting from a sabotage attack should include an association of these potential consequences with the corresponding requirements and measures. The following is an example of a three tier categorization system that associates radiological consequences with physical protection requirements:

(a) For radiological consequences that are considered to be below the State's definition of unacceptable radiological consequences, prudent management practices should be applied.
(b) For radiological consequences that are considered to be greater than unacceptable radiological consequences, but lower than the State's

definition of high radiological consequences: "the *operator* should identify equipment, systems or devices, or *nuclear material*, the *sabotage* of which could directly or indirectly lead to this condition as potential *sabotage* targets and protect them in accordance with the…design process…and protection requirements" [3].

(c) For consequences that are considered to be greater than high radiological consequences, in addition to those measures applied for consequences greater than unacceptable radiological consequences: "a minimum set of equipment, systems or devices needed to prevent high radiological consequences, should be located within one or more *vital areas*, located inside a *protected area*" [3]. Facilities such as nuclear power plants or those with comparable material inventories are more likely to be included in this category (i.e. high consequence facilities).

2.14. In accordance with the application of a graded approach, States that have a large number of nuclear facilities can define more tiers than those proposed in para. 2.13 to categorize potential radiological consequences covering the range between unacceptable and high radiological consequences. This enables the operator to implement physical protection measures that more adequately correspond to the risk for each level of potential radiological consequences, while also contributing to a more effective allocation of resources. Additional tiers could, for example, account for differences in material inventories or the attractiveness of material at various facilities to adversaries.

2.15. A graded approach should be followed, as needed, to appropriately reduce any identified security risks from sabotage. Facility operators should identify SSCs, associated operator actions, and nuclear and other radioactive material that, if sabotaged, could directly or indirectly lead to unacceptable radiological consequences. These SSCs, associated operator actions, and nuclear and other radioactive material should then be identified as potential sabotage targets and protected accordingly.

2.16. In accordance with the application of a graded approach, the stringency of the physical protection measures for nuclear security put into place for individual SSCs, the associated operator actions, and measures for nuclear and other radioactive material will depend on the potential radiological consequences. The level of physical protection for individual SSCs, associated operator actions, or nuclear and other radioactive material should be defined on the basis of the radioactive material inventory and the potential radiological consequences. In addition, the operator may choose more stringent protection for SSCs and associated operator actions that, if sabotaged, would immediately lead to unacceptable or high radiological consequences, in contrast to those that, if sabotaged, would constitute only one step in a series of sabotage acts that would be necessary to initiate unacceptable or high radiological consequences.

2.17. Security systems as a whole are designed using the principle of defence in depth. Reference [3] states:

> **"The State's requirements for physical protection should reflect a concept of several layers and methods of protection (structural, other technical, personnel and organizational) that have to be overcome or circumvented by an adversary in order to achieve his objectives."**

Further, security measures should delay an adversary's ability to accomplish an act of sabotage to allow time for additional resources to arrive to mitigate the sabotage event. Even the best security measures are effective only against finite adversary capabilities, which are typically limited by the design basis threat, as defined by the State. Security measures might ultimately fail during the course of an attack that exceeds the capabilities defined in the design basis threat or representative threat statement. In this case, the existing safety margins of the nuclear safety systems of a facility may nevertheless prevent unacceptable or high radiological consequences.

2.18. Specific criteria should be established to assess the risks associated with the threats defined in the design basis threat or representative threat statement that could result in unacceptable or high radiological consequences. The following are sample questions that could be used by the competent authority and/or the operator when assessing the risks:

(a) Is the nuclear facility sufficiently robust to prevent the immediate, uncontrolled release of significant amounts of fission products (i.e. robust against catastrophic failure) in the event of a sabotage attack?
(b) Would the essential safety systems and the necessary operator actions continue to perform their functions in the event of a sabotage attack, or could they be started and operated as needed for essential cooling, control of reactivity and containment of radioactive substances?
(c) Following a sabotage attack, could essential safety systems be operated until repairs could be carried out, even when subjected to the related effects of an attack, such as fire, smoke or structural damage?
(d) Would the facility, system design and operation, in addition to the response procedures and capabilities, adequately mitigate and minimize any exposure

to radiation of the public and facility personnel in the event of a nuclear accident caused by a sabotage attack?

IDENTIFICATION OF SABOTAGE TARGETS AND VITAL AREAS

2.19. The process described in this publication for the identification of sabotage targets can be used for the evaluation of all types of nuclear facility and throughout the lifetime of the facilities, from design to operation and ultimately to decommissioning. It can also be used for the evaluation of situations in which the assumptions used in the design basis threats, representative threat statement or sabotage attack scenarios have changed.

2.20. After the State has defined thresholds for unacceptable radiological consequences and high radiological consequences — according to the nuclear facility type and radioactive material inventory — the first step in the process should be to conduct an analysis, validated by the competent authority, to determine whether the radioactive material inventory has the potential to result in radiological consequences exceeding the thresholds for unacceptable radiological consequences or high radiological consequences. This analysis should consider only the inventory of the facility and should assume that sabotage acts will be successfully completed, without accounting for any physical protection or mitigation measures. If it is determined that there is no credible potential for any sabotage act to exceed the thresholds for unacceptable radiological consequences or high radiological consequences, no further analysis is necessary.

2.21. For type 1 sabotage threat scenarios that could result in situations with high radiological consequences or greater (as defined in paras 2.4–2.10), a process should be used to identify sabotage targets and to appropriately assign vital areas within the physical protection system to protect these targets. The identification process, which is outlined in Sections 3–6, involves the following steps:

(1) Gathering the input data needed for the analysis, which include the results of a determination by the State of the thresholds for unacceptable radiological consequences and high radiological consequences, information on facility lifetime stages and facility states to be included in the target identification process, characteristics of the threat(s), and site and facility characteristics;
(2) Identifying potential sabotage targets and the areas in which these targets are located or co-located, and performing an analysis using a logic model to determine candidate sets of areas as potential vital areas, which will

help to prevent all possible sabotage attacks from leading to unacceptable radiological consequences or greater;

(3) Developing sabotage attack scenarios on the basis of a range of possible initiating events that would deliberately be caused by an adversary in an attempted sabotage of the facility;
(4) Performing an analysis to determine which of the candidate area sets should be selected as the final vital area set, including through an assessment of the feasibility of protecting the areas contained in each candidate area set;
(5) Performing a vulnerability evaluation to assess whether the selected vital area set is sufficient.

2.22. If the design basis threat capabilities defined by the State are sufficient to cause a type 2 sabotage threat, it should be analysed whether the protection of sabotage targets in vital areas — which have been assigned using the process described in para. 2.21 — is sufficient against the potential effects of such sabotage threat scenarios. Section 7 of this publication outlines an additional process and set of considerations, along with guidance to enable a detailed engineering evaluation. The process involves the following steps:

(1) Gathering the input data needed for analysis;
(2) Developing sabotage attack scenarios based on a range of possible initiating events that would deliberately be caused by an adversary in an attempted sabotage of the facility;
(3) Undertaking a sabotage margin assessment to evaluate the capacity of safety features to resist the identified sabotage attack scenarios, including identification of one or more sabotage attack scenarios in which the facility is able to be safely shut down and maintained in a safe shutdown condition in response to the relevant sabotage attack;
(4) Considering methods for successfully mitigating the adverse consequences resulting from the vulnerabilities of SSCs important to safety or SSCs with low margins to approved safety limits for the adverse consequences under consideration.

2.23. The results of sabotage target identification provide a basis for decisions relating to safety design, including design modifications, technical measures, procedures, the assignment of responsibilities and risk acceptance criteria. IAEA Safety Standards Series Nos SSG-5 (Rev. 1), Safety of Conversion Facilities and Uranium Enrichment Facilities [15]; SSG-6 (Rev. 1), Safety of Uranium Fuel Fabrication Facilities [16]; SSG-15 (Rev. 1), Storage of Spent Nuclear Fuel [17]; SSG-22 (Rev. 1), Use of a Graded Approach in the Application of the Safety Requirements for Research Reactors [18]; SSG-42 (Rev. 1), Safety of Nuclear

Fuel Reprocessing Facilities [19]; SSR-2/1 (Rev. 1), Safety of Nuclear Power Plants: Design [20]; SSR-3, Safety of Research Reactors [21]; and SSR-4, Safety of Nuclear Fuel Cycle Facilities [22] and the results of safety analyses can provide useful information for sabotage target identification and for the evaluation of potential radiological consequences for several types of nuclear facilities during their entire lifetime.

2.24. To provide defence in depth against sabotage attacks, operating personnel and safety specialists should work in close cooperation with security specialists, the agencies responsible for emergency preparedness and response, and other competent authorities for the application of the processes for the identification and categorization of sabotage targets and the identification of vital areas in nuclear facilities.

2.25. Identification of sabotage targets and vital areas during the design stage of a nuclear facility provides an opportunity to reduce the number of sabotage targets and the size of vital areas. It also ensures the implementation of a more cost effective physical protection system. Locating SSCs that are not important to safety outside vital areas, for example, can dramatically reduce the number of personnel who need access to vital areas. Moreover, safety measures can be designed to provide better capabilities in order to mitigate initiating events of malicious origin; for instance, by placing redundant and diverse SSCs important to safety in different vital areas, thus making it more difficult for an adversary to defeat safety mitigation capabilities.

3. INPUT DATA FOR THE SABOTAGE TARGET IDENTIFICATION PROCESS

3.1. The first step in the process of sabotage target and vital area identification, regardless of the type of sabotage threat scenario, is gathering input data. This data collection process may involve cooperation between the competent authority and the operator. The operator should determine the need for performing each type of analysis depending on the State's design basis threat or the representative threat statement and the potential radiological consequences for the facility for which the sabotage assessment is being performed.

3.2. The process for the identification and evaluation of potential sabotage attack scenarios and targets may include input related to the following:

(a) The determination by the State of unacceptable and high radiological consequences (see paras 3.3–3.5);
(b) The identification of potential targets (see paras 3.6–3.9);
(c) The determination of facility states to be assessed in the target identification process (see paras 3.10–3.27);
(d) The threat characteristics (see paras 3.28–3.31);
(e) The site and facility characteristics (see paras 3.32 and 3.33).

DETERMINATION BY THE STATE OF UNACCEPTABLE AND HIGH RADIOLOGICAL CONSEQUENCES

3.3. For nuclear facilities, the radiological consequences of concern are typically the release of radioactive substances to the environment, radioactive contamination of equipment or personnel, and radiation exposure of personnel or the public.

3.4. Unacceptable radiological consequences are defined in Ref. [3] as "A level of radiological consequences, established by the State, above which the implementation of *physical protection measures* is warranted." This level is typically defined in terms of the dose, and the amount and type of radioactive material released.

3.5. Safety documentation may serve as a basis for defining the thresholds for unacceptable radiological consequences and high radiological consequences and, if appropriate, for defining categories between these thresholds. If the determination, for nuclear security purposes, of thresholds for unacceptable and high radiological consequences proves them to be identical to the thresholds that have been defined by the State in relation to nuclear and radiation safety considerations, the safety analyses performed for the facility could be used for vital area identification, although these analyses may need to be modified to consider intentionally triggered events.

IDENTIFICATION OF POTENTIAL TARGETS

3.6. Paragraphs 5.4 and 5.5 of Ref. [3] state:

> "For each *nuclear facility*, an analysis, validated by the *competent authority*, should be performed to determine whether the radioactive inventory has the potential to result in *unacceptable radiological consequences* as determined by the State, assuming that the *sabotage acts* will be successfully completed while ignoring the impact of the physical protection or mitigation measures.
>
> "On the basis of these analyses, the State should consider the range of radiological consequences that can be associated with all its *nuclear facilities* and should appropriately grade the radiological consequences that exceed its limits for *unacceptable radiological consequences* in order to assign appropriate levels of protection."

3.7. Paragraph 3.97 of Ref. [4] states:

> "The factors that should be taken into account when determining whether or not unacceptable radiological consequences are possible at a facility include the characteristics described below (as applicable):
>
> (a) The amount, type, physical form and status of radioactive material at the nuclear facility (e.g. solid or liquid form, in process or storage).
> (b) The intrinsic risk (e.g. of criticality) associated with the physical processes and chemical processes that normally take place at the nuclear facility.
> (c) The characteristics of processes or engineering features that may become unstable during an attack.
> (d) The thermal power capacity of the facility and the irradiation history of the nuclear fuel (for a nuclear reactor).
> (e) The configuration of the nuclear facility for different types of activity.
> (f) The spatial distribution of radioactive material at the nuclear facility. For example, in research reactor facilities, most of the radioactive inventory is typically in the reactor core and the fuel storage pool; in processing and storage facilities, the radioactive inventory may be distributed across the site.
> (g) The characteristics of the nuclear facility relevant to the consequences of dispersal of radionuclides to the atmosphere and the hydrosphere (e.g. the size, design and construction of the facility or the demographics and land and water features of the region).

(h) The potential for off-site versus on-site radiological contamination (which will depend in part on the location of the radioactive material relative to the site boundaries)."

3.8. Paragraph 5.20 of Ref. [3] states:

> "*Nuclear material* in an amount which if dispersed could lead to high radiological consequences and a minimum set of equipment, systems or devices needed to prevent high radiological consequences, should be located within one or more *vital areas*, located inside a *protected area*."

Paragraph 3.92 of Ref. [4] states that "States should also define the threshold for high radiological consequences, above which it is recommended that vital areas are identified and protected at a higher level".

3.9. The safety assessment of an operating facility can support the identification of potential targets. The potential radiological impacts of accidents involving spent fuel should also be assessed by the operating organization and reviewed by the competent authority. Such safety assessments can provide important input for the design and evaluation of protection against sabotage. Paragraph 5.23 of Ref. [17] states:

> "The safety assessment [for spent fuel storage facilities] should include an assessment of hazards in operational states and accident conditions. It should provide an assessment of doses at the site boundary and of the potential for exposure in areas within the site to which there is to be unrestricted access."

DETERMINATION OF FACILITY STATES TO BE ASSESSED

3.10. Facility states include operational states and accident conditions. These facility states may rely on different SSCs to perform the necessary safety functions, and distinct nuclear security measures may be needed to protect the SSCs and the nuclear and other radioactive material in the facility under the various facility states. Each facility state should be analysed to identify related sabotage targets and vital areas. Alternatively, a bounding facility state could be identified, which, if adequately protected, would ensure protection during all facility states.

3.11. Paragraphs 3.12–3.25 address considerations related to the determination of facility states that are to be considered as part of the analysis to identify sabotage targets and associated vital areas. These considerations include defence in depth for safety, fundamental safety functions, SSC unavailability and associated operator actions.

Safe states and defence in depth for safety

3.12. The analysis to identify sabotage targets and associated vital areas should be done for safe states. In the IAEA Nuclear Safety and Security Glossary [23], a safe state is defined as a "*Plant state*, following an *anticipated operational occurrence* or *accident conditions*, in which the reactor is subcritical and the *fundamental safety functions* can be ensured and maintained stable for a long time." The concept of facility states as it is used in the safety standards for research reactors and for nuclear fuel cycle facilities is broadly equivalent to the concept of plant states for nuclear power plants.

3.13. The defined safe states may differ according to the different stages of the lifetime of a facility (e.g. operation, decommissioning).

3.14. Paragraph 2.13 of SSR-2/1 (Rev. 1) [20] states (footnotes omitted):

> "There are five levels of defence:
>
> (1) The purpose of the first level of defence is to prevent deviations from normal operation and the failure of items important to safety. This leads to requirements that the plant be soundly and conservatively sited, designed, constructed, maintained and operated in accordance with quality management and appropriate and proven engineering practices. To meet these objectives, careful attention is paid to the selection of appropriate design codes and materials, and to the quality control of the manufacture of components and construction of the plant, as well as to its commissioning. Design options that reduce the potential for internal hazards contribute to the prevention of accidents at this level of defence. Attention is also paid to the processes and procedures involved in design, manufacture, construction, and in-service inspection, maintenance and testing, to the ease of access for these activities, and to the way the plant is operated and to how operating experience is utilized. This process is supported by a detailed analysis that determines the requirements for operation and maintenance of the plant and the requirements for quality management for operational and maintenance practices.
>
> (2) The purpose of the second level of defence is to detect and control deviations from normal operational states in order to prevent anticipated

operational occurrences at the plant from escalating to accident conditions. This is in recognition of the fact that postulated initiating events are likely to occur over the operating lifetime of a nuclear power plant, despite the care taken to prevent them. This second level of defence necessitates the provision of specific systems and features in the design, the confirmation of their effectiveness through safety analysis, and the establishment of operating procedures to prevent such initiating events, or otherwise to minimize their consequences, and to return the plant to a safe state.

(3) For the third level of defence, it is assumed that, although very unlikely, the escalation of certain anticipated operational occurrences or postulated initiating events might not be controlled at a preceding level and that an accident could develop. In the design of the plant, such accidents are postulated to occur. This leads to the requirement that inherent and/or engineered safety features, safety systems and procedures be capable of preventing damage to the reactor core or preventing radioactive releases requiring off-site protective actions and returning the plant to a safe state.

(4) The purpose of the fourth level of defence is to mitigate the consequences of accidents that result from failure of the third level of defence in depth. This is achieved by preventing the progression of such accidents and mitigating the consequences of a severe accident. The safety objective in the case of a severe accident is that only protective actions that are limited in terms of lengths of time and areas of application would be necessary and that off-site contamination would be avoided or minimized. Event sequences that would lead to an early radioactive release or a large radioactive release are required to be 'practically eliminated'.

(5) The purpose of the fifth and final level of defence is to mitigate the radiological consequences of radioactive releases that could potentially result from accidents. This requires the provision of adequately equipped emergency response facilities and emergency plans and emergency procedures for on-site and off-site emergency response."

3.15. Paragraph 2.12 of SSR-3 [21] provides a description of the five levels of defence in research reactors.

3.16. The purpose of each level of defence in depth for safety can be used to develop facility specific objectives for protection against sabotage when designing the facility's physical protection system. A process for designing a physical protection system, including against sabotage, is described in section 4 of Ref. [4].[4]

3.17. For example, the purpose of the third level of defence could be used as an objective for protection against sabotage, as effective sabotage protection in line with the third level would also meet the purpose of the fourth and the fifth level.

3.18. In addition, levels of defence in depth for safety could be used as part of a graded approach to categorizing sabotage attack scenarios developed during the process of identifying sabotage targets (see Section 5).

Fundamental safety functions

3.19. Requirement 4 of SSR-2/1 (Rev. 1) [20] states:

> **"Fulfilment of the following fundamental safety functions for a nuclear power plant shall be ensured for all plant states: (i) control of reactivity; (ii) removal of heat from the reactor and from the fuel store; and (iii) confinement of radioactive material, shielding against radiation and control of planned radioactive releases, as well as limitation of accidental radioactive releases."**

3.20. The fundamental safety functions for a research reactor facility, which are the same as above, are established in Requirement 7 of SSR-3 [21].

3.21. Safety analysis reports should be considered as important input in the sabotage target identification process. Paragraph 3.3.4 of IAEA Safety Standards Series No. SSG-61, Format and Content of the Safety Analysis Report for Nuclear Power Plants [24], states that the safety analysis report should identify "the plant specific safety functions that are necessary to fulfil the main safety functions and how their fulfilment is ensured by the plant's inherent features".

3.22. Other information to be provided in the safety analysis report is described in para. 3.19.1 of Ref. [24] as follows:

> "[I]nformation on emergency arrangements, demonstrating in a reasonable manner that, in a nuclear or radiological emergency, all actions necessary for the protection of workers (including emergency workers), the public and

---

[4] Phase 1 of the development of a physical protection system involves "identifying the requirements for a physical protection system" [4].

the environment could be taken, and that the decision making process for the implementation of these actions would be timely, disciplined, coordinated and effective."

3.23. Information provided in the safety analysis report, particularly on the fulfilment of the safety functions and on emergency arrangements, should be taken into consideration to determine the facility states that should be assessed as part of the analysis to identify sabotage targets and associated vital areas.

Unavailability of structures, systems and components

3.24. Unavailability of SSCs could conceivably occur concurrently with a criminal or other intentional unauthorized act, either by chance or as a result of repairs or maintenance outages. The identification of potential sabotage targets and associated vital areas should take into account all lifetime stages and facility states.

3.25. Alternative on-site arrangements to compensate for unavailability of SSCs can be arranged. For example, a mobile power system could supply cooling pump operations and provide electric power to control the reactor vessel pressure in the case of station blackout. If alternative arrangements, such as the deployment of mobile equipment or emergency response measures, are proposed, consideration should be given to the time needed for such arrangements, the possibility to implement such provisions despite the presence of adversaries and the situation in which these actions should be undertaken. No credit should be given to mobile equipment or repair work in the assessment of an emergency until the successful installation and start of the mobile equipment or completeness of the repair work has been confirmed. In some cases, a deployment delay or the unavailability of these measures may mean that they cannot be put to use in the time available to prevent high radiological consequences.

Operator actions

3.26. Safety and other analyses used as input for sabotage target and associated vital area identification frequently contain explicit or implicit assumptions about operator actions associated with SSCs. These operator actions may involve routine, emergency or accident management actions that are needed to maintain the facility in a safe state. They may also be implicit in the way that the facility response to events is modelled. During the sabotage target and vital area identification process, all the implicit and explicit assumptions about operator actions included in the safety analysis and in other analyses that may be used as input should be carefully identified to include threats to security.

3.27. After these operator actions have been identified, a decision should be made on whether they should be considered in the analysis as part of the facility response to sabotage. The availability of SSCs and operating personnel would need to be evaluated in this context, and possible recovery actions to compensate for disabled SSCs would need to be identified (see paras 3.24 and 3.25). Whether such recovery actions are to be considered as part of the facility response to sabotage should also be determined. In addition, consideration should be given to whether the sabotage attack scenario will enable the performance of these operator actions with or without additional security measures. The rationale for considering operator actions, including recovery actions, should be documented.

THREAT CHARACTERISTICS

3.28. Fundamental Principle G of Ref. [2] states that "The State's physical protection should be based on the State's current evaluation of the threat." Further, Fundamental Principle H states that "Physical protection requirements should be based on a graded approach, taking into account the current evaluation of the threat".

3.29. Paragraph 3.10 of Ref. [3] states:

> "The State should define requirements — based on the *threat assessment* or *design basis threat* — for the physical protection of *nuclear material* in use, in storage, and during *transport*, and for *nuclear facilities* depending on the associated consequences of either *unauthorized removal* or *sabotage*."

3.30. Paragraph 2.9 of Ref. [5] states:

> "An assessment of the current threat related to nuclear security, provided in threat statements such as design basis threats and representative threat statements, can be used to facilitate a risk informed approach to nuclear security and risk management at individual facilities and activities."

3.31. Paragraph 7.2 of Ref. [5] states that "In a performance based regulatory approach, design basis threats and the State's nuclear security objectives provide the basis for designing, implementing and evaluating nuclear security systems and measures." Additional information applicable to sabotage threats can be found in paras 7.3 and 7.4 of Ref. [5], including on the development of credible sabotage attack scenarios using the threat characteristics defined in the design basis threat. These credible sabotage attack scenarios should be used in the target identification and vital area identification process (see Sections 4 and 5 of this publication).

SITE CHARACTERISTICS AND FACILITY CHARACTERISTICS

3.32. Site characteristics, including population density, land use in the vicinity of the facility, high grounds or other relevant places for stand-off attacks, as well as environmental conditions, are important factors when evaluating the consequences of a potential radiological release.

3.33. Facility characteristics provide essential input for the process of identifying sabotage targets and vital areas. The characteristics that should be identified and provided for the analysis include the following:

(a) Locations, characteristics and quantities of nuclear and other radioactive material;
(b) For nuclear material: element, enrichment, quantity, physical and chemical form, and whether it is irradiated;
(c) For radioactive material: radionuclide and quantity (in terms of activity);
(d) Information about the nuclear facility's safety provisions, such as those related to shielding, criticality control, cooling, confinement, fire prevention and structural integrity;
(e) Detailed design information on processes and safety systems.

Design information on processes and safety systems, in particular, should be used to identify SSCs and associated operator actions that should be protected against sabotage.

4. DEVELOPING SABOTAGE ATTACK SCENARIOS

4.1. To determine the areas that should be protected against criminal or other intentional unauthorized acts that could lead to unacceptable radiological consequences or high radiological consequences, sabotage attack scenarios should be identified that could potentially lead to consequences greater than the unacceptable radiological consequences as defined by the State. All credible sabotage attack scenarios that are consistent with the design basis threat or representative threat statement should be considered. The criteria taken into account for the selection of these scenarios should be documented and justified. The resulting sabotage attack scenarios should be used as input for the sabotage logic model, which can be in the form of a statement, an algebraic expression or a graphical representation (see Section 5) and which is used for the identification of sabotage targets as candidates for protection in vital areas.

4.2. When developing sabotage attack scenarios, sabotage resulting from direct, semi-direct or indirect acts should be considered. Acts of direct sabotage might lead to the dispersal of radioactive material through the application of energy from an external source (e.g. an explosive or incendiary device). Semi-direct sabotage includes criminal or other intentional unauthorized acts involving explosives or other sources of energy that are used for breaching safety barriers or that could cause safety barriers to fail, resulting in radioactive material being dispersed in a manner that is usually not addressed in the safety analysis.

4.3. Acts of indirect sabotage might lead to the dispersal of radioactive material or to exposure to radiation using the energy contained within the nuclear or radioactive material, or the energy that is generated during the processing of material. Indirect sabotage attacks might not involve an adversary gaining access into the area in which the material is located, but it might involve an attack on the SSCs or associated operator actions that maintain the facility in a safe state. For example, a sabotage attack directed at the SSCs responsible for maintaining core temperature within operational limits at a nuclear power plant could result in the dispersal of radioactive material.

DIRECT AND SEMI-DIRECT ACTS OF SABOTAGE

4.4. Paragraph 3.72 of Ref. [4] states:

> "The State should consider how to protect nuclear facilities while taking into account the potential for sabotage to cause unacceptable radiological consequences. The State should also ensure that protection measures are required for the targets within the facilities which if subject to sabotage would produce such consequences."

If the release of radioactive material would lead to consequences greater than high radiological consequences, direct dispersal of the inventory should be included in the sabotage logic model (see Section 5), and the remaining steps of the vital area identification process should be performed for the inventory.

4.5. Inventories of radioactive material that might be the source of a direct release leading to unacceptable or high radiological consequences should be considered. Such inventories could include fresh nuclear fuel, irradiated fuel, radioactive sources or radioactive waste. If the potential radiological consequences of the release of radioactive material caused by the semi-direct sabotage of these materials are equal to or greater than the established unacceptable radiological consequences, relevant SSCs and associated operator actions should be considered as sabotage targets.

INDIRECT ACTS OF SABOTAGE: IDENTIFYING INITIATING EVENTS OF MALICIOUS ORIGIN

4.6. An initiating event that is deliberately caused by an adversary in the attempted sabotage of a facility is called an initiating event of malicious origin. It is important to produce a list of sabotage attack scenarios representing attacks that could trigger initiating events of malicious origin identified as potentially leading to unacceptable or high radiological consequences. The list of initiating events of malicious origin, either alone or in combination with mitigating system failures, should be included in the sabotage logic model (see Section 5), which sets out a methodology for the identification of sabotage targets.

4.7. The following two types of indirect acts of sabotage should be considered:

(a) Attacks causing an initiating event that could create conditions more severe than the facility's mitigating systems can accommodate (i.e. events that are beyond the safety design basis);
(b) Attacks causing an initiating event and disabling the systems needed to mitigate the effects of the initiating event.

4.8. Many initiating events will have already been identified and analysed in the safety documentation of the facility (e.g. in safety case studies, deterministic safety analyses, probabilistic safety assessment reports). These initiating events should be considered as potential initiating events of malicious origin.

4.9. Facility safety assessments, reports and analyses, in particular, contain valuable information and models. A deterministic safety analysis or a probabilistic safety assessment for the facility, for example, provides analyses of the response of the facility to various initiating events (see IAEA Safety Standards Series No. GSR Part 4 (Rev. 1), Safety Assessment for Facilities and Activities [25]). Recommendations on the systematic identification of hazards and accident scenarios associated with facility states and accident conditions are provided in IAEA Safety Standards Series No. SSG-68, Design of Nuclear Installations Against External Events Excluding Earthquakes [26].

4.10. Three categories of potential initiating events should be considered when identifying potential initiating events of malicious origin that might not have been included in the safety case, as follows:

(a) Initiating events in which there is no process energy or other energy source present that could lead to the dispersion of radioactive material. Such events involve criminal or other intentional unauthorized acts using explosives or other sources of energy for breaching barriers, or radioactive material being dispersed in a manner that would not be possible without a criminal or other intentional unauthorized act.
(b) Initiating events that are so unlikely to occur randomly that they are excluded from safety consideration. Such events involve multiple independent initiating events or massive breaches or failures of passive components that, while extremely improbable as random events, can be accomplished by an adversary equipped with explosives or other resources, including in situ resources. Initiating events that are considered independent from a safety point of view may sometimes be triggered by the same criminal or other intentional unauthorized attack and thus cannot be considered as independent from a security point of view. For example, the simultaneous loss of all trains of equipment, which might be the result of an explosion or several coordinated sabotage actions, would challenge the performance of safety functions beyond design.
(c) Initiating events that involve sources of release of radioactive material that might not have been within the scope of the safety analysis. For example, Level 1 probabilistic safety assessment at nuclear power plants address only events with the potential to lead to core damage and, thereby, the release of radioactive material from the reactor core. Other inventories of radioactive material that might be a source of release leading to high radiological consequences (e.g. irradiated fuel, radioactive waste) should also be considered.

4.11. There are four approaches that can be used to identify initiating events of malicious origin:

(a) Review of safety documentation. This review should be the starting point for understanding the potential hazards in a facility. The review of safety reports, assessments, deterministic safety analyses, probabilistic safety assessments, fire analyses, seismic analyses and other safety evaluations for the facility and for similar facilities will help to identify initiating events. Any of the initiating events that can occur randomly can also be caused by criminal or other intentional unauthorized acts, so this set of initiating events should be included in the list of initiating events of malicious origin. Assumptions, in particular those that might be made in safety analyses regarding the nature of initiating events and the facility's response to them, should be re-examined in the context of criminal or other intentional unauthorized acts, and should be revised where appropriate.
(b) Review of other vital area identification analyses. Where other vital area identification analyses have been performed for similar facilities, lists of the initiating events of malicious origin used for these analyses should be reviewed. Initiating events of malicious origin that do not correspond to initiating events that have been identified in facility safety documentation should be identified.
(c) Engineering evaluation. The operational and safety systems of the facility and major structures and components should be systematically reviewed to identify any additional initiating events of malicious origin. For example, any initiating events that could result from criminal or other intentional unauthorized acts in which the potential adversary is deemed capable of initiating an event that leads directly, or in combination with other criminal or intentional unauthorized acts, to unacceptable or high radiological consequences should be identified.
(d) Deductive analysis. Unacceptable radiological consequences should be systematically decomposed into postulated events that might cause them to occur. The successful operation of systems and other actions that could prevent these events are not included in this analysis. Events that might, at the most fundamental level, trigger successive events leading to unacceptable or high radiological consequences are then considered candidates for the list of initiating events of malicious origin identified for the facility.

4.12. Because the objective is to produce a list of initiating events of malicious origin that is as complete as possible, all the approaches presented in para. 4.11 should be used. Each initiating event of malicious origin should also be assessed to determine whether there are systems capable of mitigating the event. These initiating events of malicious origin will either exceed the mitigating system capability or be within the mitigating system capability (see paras 4.13–4.23).

Initiating events of malicious origin that exceed the mitigating system capability

4.13. Every initiating event of malicious origin that exceeds the mitigating system capability should be included in the sabotage logic model. An example is a pipe break that creates a loss of coolant that cannot be compensated by an injection system.

4.14. The likelihood that a given threat could cause an initiating event of malicious origin that would exceed the mitigating system capability is addressed when considering the credibility of threat characteristics later in the process (see paras 4.24–4.26).

Initiating events of malicious origin that are within the mitigating system capability

4.15. While an initiating event of malicious origin may be within the capability of the mitigating systems, if such systems are disabled, including through associated operator actions, the event might proceed unhindered. Therefore, even in the case of initiating events of malicious origin that are within the mitigating system capability, combinations of initiating events of malicious origin and of events that might disable the mitigating systems (i.e. mitigating system disablement events) and could lead to unacceptable or high radiological consequences should be determined. Such combinations of mitigating system disablement events and associated initiating events of malicious origin should be included in the sabotage logic model. For example, the loss of off-site power in combination with a loss of emergency power supply could disable sufficient decay heat removal and lead to unacceptable radiological consequences, and this combination should therefore be included in the sabotage logic model.

4.16. The credibility of the threat causing these mitigating system disablement events and initiating events of malicious origin is also addressed when the credibility of threat characteristics is considered later in the process (see paras 4.24–4.26).

4.17. The specific systems that are used to mitigate initiating events may differ depending on the facility state. Systems that are used to mitigate initiating events also support safety functions, such as reactivity control, decay heat removal, coolant boundary integrity and containment integrity. The systems that directly perform fundamental safety functions (see paras 3.19 and 3.20) are considered to be safety systems, while the equipment needed for the proper functioning of safety systems is referred to as support features. The successful operation of a safety system may depend on the availability of one or more support features. Any dependencies of safety systems on support features should be identified during the assessment of initiating events of malicious origin and mitigating system disablement events, since they could represent vulnerabilities. For example, the operation of a high pressure emergency core cooling pump needs, as a support feature, cooling of its essential components.

4.18. If a probabilistic safety assessment has been prepared for the facility, information on safety systems and support features should be readily available from this analysis or from supporting documentation. If only a deterministic safety analysis is available, most or all of this information can be derived from the accident analyses by employing engineering judgement.

4.19. If the deterministic safety analysis lists safety SSCs, the list could be helpful in identifying safety systems and their dependencies. However, there may be other dependencies beyond the safety analysis that relate to specific sabotage attack scenarios. As an example, if an installed drainage system is sabotaged, explosive breaching of a cooling water pipe might cause flooding of equipment near or below the pipe breach, regardless of whether the drainage system satisfies safety concerns on flooding as a non-malicious initiating event.

4.20. In the context of the analysis of initiating events of malicious origin, the successful operation of a safety system means that the system's safety function is performed to at least the minimum level needed to address the conditions that could be created by an initiating event of malicious origin. Information that is considered relevant to the development of success criteria for safety systems and support features is provided in the facility safety analyses. The success criteria for safety systems are of particular importance when analysing initiating events of malicious origin because they define the starting points for the subsequent sabotage logic modelling of the systems. Success criteria could include performance measures (e.g. flow rates, response times) and hardware requirements (e.g. the number of required flow paths or power trains).

4.21. Defining success criteria for support features can be more complicated than defining those for safety systems. In most cases, support features serve more than one safety system and, as a result, each possible state of the system (e.g. three trains operating, two trains operating, one train operating, no train operating) has a different effect on the safety systems that perform the particular safety function. The success criteria for a support feature vary depending on the different safety functions and the associated safety systems.

4.22. For some facilities, there might be a large number of potential initiating events of malicious origin that, coupled with mitigating system disablement events, could lead to unacceptable radiological consequences or greater. For such facilities, it may be better to group together events for which the same safety system and support feature performance is needed to mitigate the initiating event of malicious origin. Grouping events in this way reduces the complexity of the sabotage logic model. In a given group of initiating events of malicious origin, the safety systems and support features should then essentially meet the same success criteria, which would permit sabotage logic models to begin with any of the events in the group. In addition, the sabotage logic model would apply to all of the events in the group. Where only a small number of potential initiating events of malicious origin have been identified for the facility, it may be preferable not to group events.

4.23. If a probabilistic safety assessment has been performed for the facility, the documentation for this analysis should contain the grouping of initiating events considered in the analysis. The same groupings can be employed for corresponding initiating events of malicious origin. If a probabilistic safety assessment has not been performed for the facility, it may be possible to use other safety documentation or another source of information to begin with groupings of initiating events. However, event groupings depend on the design of the facility, so groupings taken from other sources should be carefully evaluated to ascertain whether they are appropriate for the facility being analysed.

CREDIBILITY OF SABOTAGE ACTS

4.24. The analysis of acts of sabotage in the preceding subsections does not consider the capability of the threat to perform these acts. All events that could lead directly or indirectly to levels of unacceptable radiological consequences or greater are initially included in the analysis to ensure that potential sabotage targets are not overlooked, but without regard to whether the defined threat capabilities and sabotage attack scenarios are sufficiently credible to result in initiating events of malicious origin. If the assumed threat characteristics change, the information developed in the preceding steps remains valid under the changed threat conditions.

4.25. Paragraph 6.10 of Ref. [5] states:

> "Targets for which malicious acts could lead to unacceptable radiological consequences, as defined by the State, should be identified. These targets should then be considered in conjunction with the attributes and characteristics of the potential adversaries described in the national nuclear security threat assessment documentation in order to identify threats that are relevant to these targets and that might therefore cause unacceptable radiological consequences."

Paragraph 6.11 of Ref. [5] states that "If the capabilities of a given adversary are not sufficient to commit such an act, then that adversary may be excluded from further consideration."

4.26. The credibility of the direct dispersal of material, of causing an initiating event of malicious origin and of disabling mitigating systems should be assessed. Events that are not credible may be removed from the sabotage logic model. Preventing, for example, the loss of off-site power is practically impossible for a facility's physical protection system. The loss of off-site power can be achieved in many ways without having to gain access to the facility. As a boundary condition for determining the credibility of sabotage attack scenarios, it should therefore be assumed that off-site power is unavailable.

5. IDENTIFICATION OF POTENTIAL SABOTAGE TARGETS

5.1. Once the inputs needed for the process of identifying and evaluating potential sabotage attack scenarios and targets have been determined, and a list of initiating events of malicious origin and associated sabotage attack scenarios has been compiled on the basis of these inputs, the next step is to identify sabotage targets that could be candidates for protection in vital areas (information on vital area identification is provided in Section 6).

5.2. The minimum set of SSCs and associated operator actions to be protected may include all safety systems, if dictated by the overall safety philosophy at the facility, or alternatively, a minimum set may be defined as a subset of all SSCs and associated operator actions. The number and extent of the SSCs and associated operator actions to be protected, along with the designated vital areas, is specific to each facility.

5.3. In order to identify sabotage targets that could be candidates for protection in vital areas, a sabotage logic model should be developed that uses as input the sabotage attack scenarios described in Section 4 concerning direct and semi-direct sabotage and initiating events of malicious origin, along with the SSCs and associated operator actions that could mitigate these initiating events (if any). The areas in the facility from which an adversary could accomplish each event in the sabotage logic model should then be identified and documented, and candidate area sets should then be identified.

SABOTAGE LOGIC MODEL

5.4. A sabotage logic model can be a statement, an algebraic expression or a graphical representation, such as a logic fault tree or a logic event tree, used to identify events or combinations of events that could lead to high radiological consequences.

5.5. The sabotage logic model should include all direct and semi-direct sabotage attacks, and all initiating events of malicious origin that exceed the mitigating system capacity[5] as single events leading to high radiological consequences. Initiating events of malicious origin within the mitigating system capacity should be included alongside associated mitigating system disablement events. The part of the logic model relevant to mitigating system disablement should be developed to the component level, using a top down approach, and in sufficient detail to enable the linking of disablement events to the facility locations (i.e. areas) in which the system disablement can be accomplished.

5.6. Information provided in the facility safety analysis and other safety documentation can be used to develop the sabotage logic model for initiating events of malicious origin within the mitigating system capacity.

---

[5] In the context of this publication, capacity refers to an 'absolute' measure of the robustness of SSCs subjected to a particular threat that can include physical, operational and administrative attributes. Capacity is defined relative to a specific metric. Code capacity is a measure of a plant design feature relative to the code. Failure capacity is a measure of the robustness of SSCs subjected to a particular threat.

5.7. The development of the sabotage logic model for initiating events of malicious origin within the mitigating system capacity is typically carried out in two steps:

(1) Development of a facility sabotage logic model representing combinations of initiating events of malicious origin and the disablement of safety systems leading to high radiological consequences, which is accomplished using the information outlined in para. 5.5 along with information from the facility safety analysis;
(2) Development of sabotage logic models for individual safety systems and the support features that they are dependent upon, either by modifying existing logic models from the facility's probabilistic safety assessment or by developing logic models using facility system configuration information, as well as success criteria and dependency information.

5.8. The second step in the process presented in para. 5.7 produces the portion of the sabotage logic model that links each initiating event of malicious origin with the disablement of the mitigating safety systems, as well as the corresponding support features and associated operator actions. An example of a simple sabotage logic model is provided in Annex I.

FACILITY WALKDOWN AND IDENTIFICATION OF AREAS

5.9. The areas surrounding SSCs and associated operator actions should be identified as candidates for the implementation of nuclear security measures. However, it would not be reasonable to design and implement specific nuclear security measures to protect individual SSCs and associated operator actions.

5.10. After developing the sabotage logic model, areas in the facility from which an adversary could accomplish each event in the sabotage logic model should be identified and documented. These areas may later be identified as candidate vital areas (see paras 5.15–5.17 and Section 6). Information about these areas should be collected through a structured process and then verified by conducting a walkdown of the facility. Annex II provides a detailed example of a facility walkdown.

5.11. Design documents for the nuclear facility provide the information needed to identify areas in which sabotage attacks can be accomplished. General arrangement drawings should depict the area, room, walls and doors, as well as information on access routes. Piping and instrumentation diagrams, isometric drawings, safe shutdown analyses, and fire, flood and seismic probabilistic safety assessments are other sources of information on equipment locations.

5.12. In preparation for the walkdown, location information for the overall facility (e.g. buildings, doors and windows, rooms or compartments) should be reviewed. The walkdown should be performed by representatives from the facility's safety, security, design and operating organizations. The main objectives of the walkdown are the following:

(a) To verify the areas from which the threat capabilities defined in the design basis threat or representative threat statement could accomplish direct dispersal of nuclear or other radioactive material.
(b) To verify the set of areas from which the threat capabilities defined in the design basis threats or representative threat statement could accomplish each identified initiating event of malicious origin.
(c) To verify the set of areas from which the threat capabilities defined in the design basis threat or representative threat statement could accomplish each of the actions needed to disable sabotage targets or associated operator actions identified in the sabotage logic model.
(d) To assess the potential for spatial interactions among adjacent areas. External event probabilistic safety assessments, such as those for seismic events, fire and flooding, may provide useful information on spatial interactions.

5.13. A vital area should be capable of supporting the detection and delay functions at its boundaries, and it should be positioned so as to enable the timely response of response forces. For each area identified, it should be feasible to either employ existing structures or create new constructions in order to establish a physical barrier that controls access to the area and minimizes the number of entrances to and exits from the area — with due consideration given to the safety aspects of emergency situations — as well as to detect unauthorized access to the area. Such an approach can provide a more effective and cost effective solution if taken into account at an early stage of the design of the nuclear facility.

5.14. All vital areas identified should be documented on facility arrangement drawings or other facility design and layout documents to clearly define the area boundaries.

SABOTAGE AREA LOGIC MODEL

5.15. After identifying the areas in the facility in which an adversary could accomplish each event in the sabotage logic model, the sabotage area logic model should be created by replacing each event in the sabotage logic model (i.e. each direct dispersal event, initiating event of malicious origin and mitigating system disablement event) with the area or areas in the nuclear facility in which an adversary could cause the event. Depending on the approach, this might be accomplished automatically through some type of linking table (e.g. a location map) or manually by modifying the sabotage logic model directly so that the initiating events under consideration are replaced by areas in which they can be performed.

5.16. The sabotage area logic model should be 'solved' in order to find a minimum set of areas accommodating targets that should be protected against sabotage and which would prevent all possible sabotage attack scenarios from leading to unacceptable radiological consequences or greater. The following steps should be followed to solve the sabotage area logic model:

(a) Identification of all combinations of areas to which an adversary might gain access to complete a sabotage attack that could lead to unacceptable radiological consequences or greater.
(b) Determination of the minimum combinations of areas that should be protected to ensure that no sabotage attack resulting in unacceptable radiological consequences or greater can be completed. In other words, at least one of the areas in each combination of areas in which sabotage can be accomplished should be protected.

5.17. Each combination of locations for which protection will prevent all sabotage attack scenarios from taking place constitutes a candidate vital area set (see Section 6). Each of the candidate vital area sets should be capable of ensuring detection, delay and response measures against sabotage acts.

6. IDENTIFICATION OF VITAL AREAS

6.1. The SSCs and associated operator actions identified as targets to be protected in vital areas within the selected vital area set should be listed in the safe shutdown equipment list, along with the corresponding operator actions. The

SSCs listed in the safe shutdown equipment list are those SSCs that should be in functioning order to safely shut down the facility and maintain the facility in a safe shutdown condition. For a nuclear power plant, this safe shutdown equipment list will contain a few hundred SSCs. Other types of facility may, however, have significantly fewer SSCs in their safe shutdown equipment lists. The operator actions to be listed alongside the SSCs in this list are those that should be completed in order to safely shut down the facility and maintain it in safe shutdown condition.

6.2. If the evaluation of SSCs and associated operator actions is being performed at a facility with nuclear security measures already implemented, it is expected that the majority of the SSCs in the safe shutdown equipment list will be located in previously identified vital areas and that the corresponding operator actions will take place in previously identified vital areas. However, vital area designations should be revisited as part of the evaluation. For sabotage attack scenarios that have not been previously considered in the assessment, some areas may need to be designated as vital; similarly, some areas previously designated as vital may no longer need to be defined as such. A re-evaluation after design changes at the facility could show similar results.

VITAL AREA SET SELECTION

6.3. Once the candidate sets of vital areas have been identified, a vital area set should then be selected. When selecting a vital area set, it is important to ensure that all sabotage targets identified from the safe shutdown equipment list are located within vital areas.

6.4. When selecting this vital area set, the operator should also account for factors that are important for the safe and efficient operation of the facility. An overly conservative approach to selecting the vital area set could potentially result in unacceptable impacts on the safety of the facility, facility operations and emergency response, and in high costs, given the increase in measures necessary to ensure the protection of identified sabotage targets. Specific needs that may have been identified for access to some areas and SSCs could influence the selection of candidate vital areas. For example, the operator may propose a candidate vital area set considering the aforementioned aspects that provides the optimum combination of the following:

(a) Lowest impact on safety, facility operations and emergency response;
(b) Lowest difficulty in terms of implementing protection measures;

(c) Highest effectiveness of protection measures;
(d) Lowest cost of protecting the vital areas.

6.5. The selected vital area set should include the following areas:

(a) All areas from which the threat has the capability to cause direct or semi-direct dispersal of nuclear or other radioactive material;
(b) All areas from which the threat could cause initiating events that exceed the mitigation capability of facility systems;
(c) All areas from which the threat could cause an initiating event, and in which either mitigating SSCs and associated operator actions are located or a minimum set of mitigating SSCs and associated operator actions are located.

6.6. As part of the vital area selection process, a table should be produced that is used to evaluate each of the candidate vital area sets in relation to each of the attributes considered in the selection of a vital area set. The table should record the aggregate score or ranking of each candidate vital area set, with the associated rationale (e.g. an adversary path was considered but not determined to be credible, because it was too complicated). A recommended vital area set should be selected on the basis of the best overall score or ranking.

6.7. It is unlikely that one candidate vital area set will receive the highest rating for each of the selection criteria. Trade-offs should thus be assessed in selecting the final vital area set, which can be accomplished using engineering judgement or a more structured analytical approach.

VULNERABILITY EVALUATION OF SABOTAGE TARGETS

6.8. After the vital area set has been selected, a vulnerability assessment should be undertaken to determine whether the nuclear security measures implemented at the selected vital area set are sufficient for protection against sabotage. In the first step of the vulnerability evaluation, the vital area set is assessed to determine whether the appropriate selection was made of the target set within the vital area set to prevent high radiological consequences. If the appropriate selection was made, a review should then be undertaken to assess whether the vital areas identified in the vital area set can support detection and ensure the appropriate delay for a timely response. This review should be carried out using the facility layout and information on response capabilities, any safety constraints and the access of facility personnel to the equipment (e.g. number and frequency of personnel accessing the equipment).

6.9. If the capability of the vital area set to support detection and delay measures is considered adequate, then a subsequent assessment should be initiated to determine whether the targets inside the vital areas, protected with detection and delay measures, are vulnerable to attack from the threats described in the design basis threat or the representative threat statement. If the targets to be protected in vital areas are assessed to be vulnerable, fortification of these vital areas should be considered, through use of the facility layout, information on response capabilities and their timeliness, safety concerns and engineering safety.

6.10. If a vital area cannot be fortified, one or several of the following actions should be considered:

(a) Strengthening of the target robustness;
(b) Implementation of contingency measures;
(c) Modification of required safety functions;
(d) Assessment of the safety margin;
(e) Refinement of the target set;
(f) Implementation of additional measures that prevent access to the target.

6.11. Depending on the adversary capabilities defined in the design basis threat or the representative threat statement, insider threats alone or in collaboration with external threats may need to be taken into consideration for the vulnerability evaluation, which could lead to the identification of the need for additional security measures (see IAEA Nuclear Security Series No. 8-G (Rev. 1), Preventive and Protective Measures Against Insider Threats [27]).

7. OFF-SITE SABOTAGE ATTACKS

7.1. A nuclear facility can be subject to both type 1 and type 2 sabotage threats, and the physical protection system should be designed and evaluated on the basis of these two separate types of threat scenario.

7.2. Type 1 sabotage threat scenarios represent scenarios involving the physical presence of adversaries on the site, while type 2 sabotage threat scenarios do not involve intrusion into the facility. The options to adjust the design of the physical protection system's functions for detection, delay and response against type 2 sabotage threats are limited and may need support from State organizations.

7.3. Given the distinction between type 1 and type 2 sabotage threat scenarios, type 2 sabotage threat scenarios often need a separate analysis involving considerations related to engineering safety aspects that concern the protection of nuclear facilities against sabotage. Sections 3–6 address the identification of sabotage targets and vital areas, with a focus on type 1 sabotage threat scenarios, whereas Section 7 focuses on special considerations and the additional analysis needed to consider sabotage attack scenarios launched from outside the facility boundary.

7.4. Considerations related to type 2 sabotage threat scenarios are provided in paras 7.5–7.15, and guidance is set out for the development of relevant facility specific type 2 sabotage threat scenarios (see paras 7.16–7.31), for a sabotage margin assessment (see paras 7.32–7.44) and for the identification of success paths for the facility against the attack scenarios (see paras 7.45–7.52). A methodology for taking decisions on how to cope with the identified vulnerabilities of SSCs important to safety is presented in paras 7.53–7.61.

CONSIDERATIONS FOR TYPE 2 SABOTAGE THREAT SCENARIOS

7.5. Paragraph 3.40 of Ref. [3] states that "The State should give attention to providing protection measures against any airborne threat and against possible *stand-off attacks* specified in the State's *threat assessment* or *design basis threat*."

7.6. Some type 2 sabotage threat scenarios might involve threats that exceed the design basis threat. Such threats would therefore be outside the capacity of the facility's physical protection system and would remain the responsibility of the State.

7.7. The various systems used to protect the facility will experience loading as a result of a sabotage attack. Extreme loads correlate to extensive pressure placed on SSCs, which could result in the failure of the SSCs to adequately mitigate the sabotage. Such situations are referred to as 'extreme environmental loads'.

7.8. In some cases, the attacks outlined in type 2 sabotage threat scenarios could result in extreme loads on the facility, and the SSC capacity to manage these loads should be evaluated. Following such an analysis, the SSCs and associated operator actions that need to be more stringently protected should be determined, given the consequences resulting from the compromise or cessation of the SSCs' operability and/or their lack of capacity to manage extreme loads. The acceptance criteria to be used by the operator in this evaluation should be agreed by the competent authority.

7.9. To support the evaluation of SSCs, an extreme environmental load matrix should be developed, compiling environmental loads and load combinations for engineering evaluations, and should include the following:

(a) The name of the SSC;
(b) The type of component (e.g. wall, pump, valve, tank, pipeline);
(c) The manufacturer;
(d) The design conditions;
(e) The function;
(f) The physical location;
(g) The environmental load conditions, such as direct or indirect impact effects, direct or indirect blast effects, heat and fire loading, vibration and the effects of smothering on operability (e.g. smoke effects from fire or flooding from an internal or external source).

In addition, extreme environmental loads and supporting data should be identified for each of the SSCs in the matrix. A suggested format for the matrix and examples of extreme environmental loads are provided in Annex III.

7.10. Paragraph 8.1 of Ref. [5] states:

> "The national nuclear security threat assessment documentation should be periodically reviewed to assess whether the assessment still represents a comprehensive and balanced view of the credible threats to nuclear security in the State, and the assessment should be revised if necessary."

If this review shows that a credible type 2 sabotage threat scenario has become an actual subject of concern, an evaluation should be undertaken.

7.11. In such cases, realistic margins should be used to re-evaluate the facility, and the results may in turn lead to the need for upgrades in the physical protection system of the facility and perhaps the inclusion of some type 2 sabotage threat scenarios in the process described in Sections 4–6 for vital area identification and physical protection system design and evaluation. Alternatively, the evaluation may result in a decision concerning the feasibility of continuing operations. The State may also decide to take on part of the responsibility for protection against type 2 sabotage threat scenarios.

7.12. In a nuclear emergency, the operator needs to be capable of overseeing and managing severe facility conditions in order to take actions to mitigate the potential consequences of a sabotage attack. The operator security plan should consider the possibility of continued adversary presence at the site (i.e. an insider) whose aim is to hinder or disrupt mitigation activities.

7.13. Off-site emergency response should be available to mitigate the off-site radiological consequences of a criminal or other intentional unauthorized act that has led, or has the potential to lead, to the loss of fission product barriers through the loss of safety functions or the compromise or inoperability of safety systems. National emergency and contingency response plans should include all actions performed by State organizations in cooperation and coordination with the operating organization to cope with such situations. These actions will include specific measures to counter criminal or other intentional unauthorized acts aimed at disrupting and disabling the emergency response (see IAEA Nuclear Security Series No. 37-G, Developing a National Framework for Managing the Response to Nuclear Security Events [28]). The response of competent authorities may involve active support of the operator's response to an attack on the facility, including emergency response actions on the part of the emergency organizations in the State.

7.14. Paragraph 3.41 of Ref. [3] states that "The State should ensure that the State's *physical protection regime* is capable of establishing and maintaining the risk of *unauthorized removal* and *sabotage* at acceptable levels through risk management." The State should decide whether, with the implementation of all the available layers of defence in depth, the remaining risk from any particular sabotage attack scenario is within the established limits of acceptability. The responsibilities and the roles of the operator and the competent authorities should be outlined in the context of this decision — including the role of security response organizations and emergency response organizations.

7.15. The consideration of type 2 sabotage threat scenarios in the design of SSCs (i.e. security by design) may improve robustness in terms of external hazards in general and could help to avoid the need for potential upgrades of the physical protection system at a later stage in the lifetime of the facility.

DEVELOPMENT AND SELECTION OF FACILITY SPECIFIC ATTACK SCENARIOS

7.16. Similar to the development of type 1 sabotage threat scenarios described in Section 4, type 2 sabotage threat scenarios should be developed on the basis of the threat detailed in the design basis threat or the representative threat statement. The scenarios should be refined to account for the specific characteristics of the facility being evaluated.

7.17. This process may lead to the exclusion of some scenarios on the basis of site characteristics and facility characteristics, the type and number of inventories of radioactive material, the type and number of other facilities at the site, the design of the facility and any off-site security measures independent of the facility.

7.18. The facility's surrounding topography and vegetation may be sufficient to exclude certain attack scenarios that could be initiated outside the facility boundary. Notably, the location and layout of the facility site may limit the likelihood that particular on-site areas will be affected by certain attack scenarios. Considering nuclear security solutions in the design stage may provide an opportunity to adapt the general layout of a facility and improve protection measures against stand-off attacks. For example, office buildings on the site may be placed astutely to prevent a potential straight path and clear sight of the facility, or the facility's location in hills, mountains or a valley may limit feasible approach angles and the speed of a large aircraft attack on the site. Other factors, such as the location of electric transmission lines, may limit approach paths for attacks by aircraft. For blast load conditions, the shielding of structures provided by topographic effects and adjoining structures may limit areas of influence, and therefore should be considered. Alternatively, containment structures that can withstand stand-off attacks can be considered as vital areas protecting sabotage targets.

7.19. Potential site characteristics that might benefit adversaries should also be taken into careful consideration; for example, the proximity of nuclear facilities to public transport infrastructure (e.g. roads, railways, airports), industrial areas or populated areas. As another example, research reactors tend to be located within research centres or on university campuses, which might make the identification of potential attackers difficult. Potential site characteristics that might hinder adversaries should also be taken into careful consideration when assessing the credible accomplishment of a scenario.

7.20. The type and number of facilities, as well as the inventories of radioactive material, on the site should also be considered, since some type 2 sabotage threat scenarios might have a simultaneous impact on all the facilities. A nuclear power plant site, for example, may contain several reactor units, which potentially have interdependent safety systems. Multiple unit sites often depend on the availability of companion unit systems when addressing non-common cause failure events. In addition, other facilities with potential high radiological consequences may be present within the facility boundary, such as those devoted to spent fuel storage in fuel pools or to dry cask storage. Further, research reactor sites may have associated laboratories, isotope production facilities and hot cells.

7.21. The evaluation of type 2 sabotage threat scenarios should thus take into consideration all on-site facilities and any interdependence of their safety systems. Such considerations should include the consequence assessment of possible environmental discharges that are cumulative for all the facilities on the site and any limits in the number of emergency response forces.

7.22. High radiological consequence nuclear facilities, including nuclear power plants or reprocessing facilities, are designed for a wide range of extreme environmental load conditions. The safety design basis incorporates measures to defend against internal and external events — such as fire, pipe whip, loss of coolant accidents, earthquakes, extreme winds, explosions or aircraft impacts — and provides an 'envelope' of protection. This protection should be taken into account when evaluating type 2 sabotage threat scenarios. In fact, some scenarios may be excluded from further consideration because they are effectively bounded by design basis conditions. Bounding can be demonstrated on the basis of the event (i.e. in relation to the entire facility), the extreme load (i.e. in relation to each SSC) or the design requirements for the SSCs derived from the loads (e.g. acceleration, temperature).

7.23. Finally, off-site security measures independent of the facility's physical protection system should also be considered during the evaluation. These measures can range from increased security in the aviation industry to surveillance performed by off-site security organizations in the vicinity of the site. If such measures are in place and are effective, the competent authority may allow the operator to exclude certain sabotage attack scenarios from consideration or may use these measures to better define the bounding conditions of scenarios.

Screening process for type 2 sabotage threat scenarios

7.24. The following two methods are generally used to determine whether a type 2 sabotage threat scenario should be further considered:

(1) Postulation of the magnitude of the potential consequences of the attack;
(2) Determination of the approximate probability of the scenario's accomplishment.

7.25. In the first method, the potential effects on the facility's safety are assessed, and if the consequences are found to be insignificant, the scenario is removed from consideration in the overall sabotage assessment. An attack scenario involving a vehicle containing explosives, for example, could be removed from consideration on the basis of the effective barrier's distance from the safety systems of interest.

7.26. In the second method, determining the probability of a scenario's accomplishment and applying probabilistic screening — in parallel to the probabilistic screening for safety, which is undertaken for scenarios describing events of accidental origin — is generally complex and uncertain in the context of physical protection. Applying this method when considering scenarios involving criminal or other intentional unauthorized acts might result in significant errors, such as omitting potential types of event or overestimating or underestimating the likelihood of a specific event. However, this method may be applied to events that have not been eliminated by the process described in para. 7.25.

7.27. Sabotage acts do not lend themselves to probabilistic screening on an absolute probability basis. Moreover, there is significant difference between the probabilistic screening of events of accidental origin and of events resulting from sabotage. In the case of the screening criteria for accidental external events, it is generally assumed that scenarios with a larger damage potential will occur with lower frequency — that is, the larger the event, the lower the frequency of occurrence. In the case of sabotage, depending on the adversary's objectives and capabilities, this assumption is not applicable. For example, a State can decide that intentional plane crashes have to be taken into consideration from a security point of view, even if the probability of accidental plane crashes is considered extremely low from a safety perspective.

7.28. Probabilistic safety assessment tools (or dynamic probabilistic safety assessment tools) that are adapted to address type 2 threat scenarios, where conditional end metrics are calculated, may be used to provide the basis for screening an individual scenario. This approach assumes that the most upstream event is deterministic (i.e. $p = 1$), but sequences evolving from this event may be represented probabilistically on the basis of the facility layout, systems design and structural robustness.

7.29. One possible approach is to use a probability level for the screening of sabotage attack scenarios that is lower than that used for accidents and extreme events. The smaller order of magnitude would be used to ensure a conservative assessment and to ensure that no event is excluded as a result of the approximate nature of the probabilistic screening procedure. However, consideration should be given to uncertainties associated with assigning probabilities to sabotage attacks, which could result in the misapplication of limited resources.

7.30. To lessen the burden involved in evaluating a facility with sabotage attack scenarios, type 2 sabotage threat scenarios could be grouped according to similarities in relation to the effects on the nuclear facility, and one scenario or a composite of the grouped attack scenarios could be selected for detailed evaluation. Grouping the scenarios in this way can reduce the overall number of scenarios being considered to a more manageable number. A panel of experts in attack scenario development and nuclear safety could be appointed for such an activity.

7.31. SSCs that have low capacity and low safety importance can be screened out.

SABOTAGE MARGIN ASSESSMENT FOR TYPE 2 SABOTAGE THREAT SCENARIOS

7.32. A sabotage margin assessment involves evaluating the capacity of safety systems to resist the attacks identified in type 2 sabotage threat scenarios. This procedure starts with the definitions of extreme environmental loads and load combinations for engineering evaluations and results in an evaluation of the SSC capacity when subjected to the extreme environmental load(s) specified.

7.33. An extreme environmental load evaluation should be undertaken to evaluate the capacity of the nuclear facility's safety systems to withstand criminal or other intentional unauthorized acts (see Annex III for an example methodology). As part of this evaluation, extreme environmental load definition matrices should be developed, which include a short description of each relevant type 2 sabotage threat scenario and the extreme environmental loads and load combinations for engineering evaluations. The extreme environmental loads in the matrix may comprise, for example, impact, explosion and blast, heat and fire, vibration, hazardous material release, flooding and other site specific conditions. An example of such a matrix is provided in Fig. III–1 in Annex III.

7.34. The overall performance criteria should then be defined for the nuclear facility subjected to extreme environmental loads. In all cases, the performance criteria, including the duration of shutdown before additional aid can be mobilized from outside the facility boundaries, should be identified and approved by the competent authorities. For example, in the case of a nuclear power plant subjected to an attack outlined in a type 2 sabotage threat scenario, the overall performance criteria may be defined as hot or cold shutdown for 24 hours after the attack is initiated, with the assumption that additional aid from outside the facility boundary can be effectively mobilized within 24 hours.

7.35. Following the establishment of performance criteria, the assumptions to be used in the engineering evaluation should be defined. Examples of assumptions for nuclear facilities are the following:

(a) Loss of off-site power;
(b) Operational activities (e.g. material transport, material movement, full power operation, shutdown, refuelling, maintenance);
(c) Engineering design criteria (e.g. redundancy of the success path(s), separation, diversity).

7.36. In the next steps, the SSC capacity criteria should be defined, with one or more success paths being identified (see paras 7.45–7.52). Once the success paths have been determined, the relevant SSCs needed should be added to the safe shutdown equipment list. For each SSC in the safe shutdown equipment list, the functional specifications for system performance success should be defined, with the extreme environmental loads for these components also included in the extreme environmental load definition matrix. The failure modes to be identified, evaluated and verified are directly related to these extreme environmental loads. The evaluation of the capacity or fragility of SSCs relies to a large extent on the combined expertise and experience of the engineering safety personnel carrying out the evaluation.

7.37. It should then be confirmed that each candidate vital area set identified in the vital area identification process contains the necessary equipment for at least one success path. An alternative approach is to determine candidate vital area sets and then perform the capacity evaluation on some or all of them.

7.38. Given the assumptions presented in para. 7.35, it is important that the SSCs comprising the success path(s) are identified. The specific functions that these SSCs need to perform during and after a sabotage attack should be set out, noting that some attack scenarios may have such a large scope of affected areas that a simple screening of the overall facility site for the likelihood of significant damage within these affected areas may increase the number of SSCs to be evaluated. SSCs within the affected areas of the attack scenario may be reasonably assumed to fail, and their further detailed consideration is thus unwarranted.

7.39. The capacity of SSCs when subject to extreme environmental loads should be evaluated. This evaluation should include the following steps:

(1) Facility familiarization, many aspects of which are acquired during the determination of success path(s), the generation of the safe shutdown equipment list and the determination of environmental loads of the safe shutdown equipment list components. Additional familiarization with specific documents for SSCs of interest could be gained during this step.
(2) In-office and in-facility evaluations of SSCs in the safe shutdown equipment list, including a facility walkdown. In-office evaluations should involve the assembling of design and qualification data for the specific SSCs in the safe shutdown equipment list. Calculations should be made as necessary to determine the environmental loads and the failure probability and capacity of the SSCs.
(3) Confirmation of assumptions made in all phases of the evaluation during the facility walkdown and documentation of the facility walkdown.
(4) Documentation of the overall SSC capacity evaluation.

7.40. In evaluating SSC capacity, considerable flexibility may be necessary to combine engineering judgement based on experience with experimental data and analysis in order to obtain the capacities of SSCs along a given success path. Careful documentation is essential to ensure that all SSCs and associated operator actions on the success path have been thoroughly evaluated and meet the environmental conditions of the attack scenario under consideration.

7.41. A facility walkdown should be performed to: review the screening that has been performed; identify new, and review proposed, easy fix concepts; review identified success paths and the SSCs in the safe shutdown equipment list; compare ‘as is’ conditions with design information; group similar SSCs and their demand environments; and review vital area definitions and boundaries. Considerations for facility walkdowns are presented in paras 5.9–5.14 and an example is given in Annex II.

7.42. The potential impacts of some type 2 sabotage threat scenarios on the safety of a nuclear facility may have some shared features with external events such as earthquakes and fires. Useful information can be found in Ref. [29] and in IAEA Safety Standards Series Nos SSG-9 (Rev. 1), Seismic Hazards in Site Evaluation for Nuclear Installations [30], and SSG-64, Protection Against Internal Hazards in the Design of Nuclear Power Plants [31].

7.43. The sabotage margin assessment for type 2 sabotage threat scenarios should be performed by the following experts:

(a) Nuclear safety experts who are knowledgeable about facility systems, security, operations, emergency response and engineering and who are responsible for converting stand-off attack scenarios into specific extreme load conditions in the different areas of the facility;
(b) Physical protection experts;
(c) Experts in facility operations and on-site emergency management;
(d) Experts in engineering safety assessments and system design and engineering, including experts in modelling external hazards (e.g. civil, structural, electrical, mechanical, instrumentation and control);
(e) Experts in areas involving missiles, aircraft attacks or demolition, where relevant;
(f) Experts in vulnerability analysis and effectiveness evaluation.

7.44. The evaluation of structures should be performed using the structural acceptance criteria provided in Ref. [29]. For evaluation purposes, less stringent acceptance criteria than the safety design criteria could be used.

IDENTIFICATION OF FACILITY SUCCESS PATHS

7.45. After the relevant type 2 sabotage threat scenarios have been identified for analysis, the next step is to identify one or more success paths for the facility, or scenarios that would allow the facility to be safely shut down and maintained in a safe shutdown condition in response to the relevant attack outlined in the type 2 threat scenario.

7.46. Criteria are used to define both what is meant by success (i.e. safe shutdown alone or with additional requirements) and the number of success paths needed. Depending on the criteria, success in response to a type 2 sabotage threat may refer only to fundamental safety functions, such as reactivity control, confinement and residual heat removal (i.e. safe shutdown) at a defined safety level.

7.47. A success path is a minimum set of systems and associated operator actions and typically does not comprise all safety systems. Several possible success paths generally exist. Each success path comprises a set of SSCs and associated operator actions, whose operability and survivability are sufficient to safely shut down the facility and maintain it in a safe shutdown condition for the period specified.

7.48. Success paths should be compatible with facility operations. In addition, the success paths should take into account facility operator training and established procedures, while recognizing that for some type 2 sabotage threat scenarios, the damage to the facility resulting from an attack might be so extensive that existing training and procedures are not applicable or adequate.

7.49. SSCs subjected to extreme environmental loads on success paths should have the capacity to withstand such loads through design or through specific measures that offer protection against sabotage.

7.50. If, in the course of the assessment, the successful performance of one or more of these fundamental safety functions cannot be demonstrated, different means of restoring and maintaining containment integrity and reducing a radioactive release should be considered.

7.51. A three tiered approach could be used to define success paths and acceptance criteria for SSC performance, as follows:

(1) The first tier could apply to type 2 sabotage threat scenarios that are not catastrophic. In this case, the evaluation criteria are similar to safety design basis considerations — that is, full system redundancy (i.e. adherence to single failure criteria and redundant paths) and SSC performance limits at design levels. Examples of such scenarios are the impact of a light aircraft on the site or a vehicle bomb explosion at some distance from the facility. In these cases, it is feasible to restart the facility after inspections have been performed.
(2) The second tier could apply to type 2 sabotage threat scenarios, where only a single success path needs to be demonstrated for all of the attack scenarios in the tier (i.e. a means to keep control over the reactor core, cool the fuel and contain the release of radioactive substances). In such cases, structure and system acceptance criteria may be significantly relaxed compared with the safety design basis, taking into account the possibility of permanent deformations of structures and components.
(3) The third tier could apply to type 2 sabotage threat scenarios involving very extreme loads that could be catastrophic — for example, the impact of a

large commercial aircraft at high speed or the impact of multiple missiles on the site. In these cases, response should include on-site and off-site emergency measures.

7.52. A scenario in any of these tiers would lead to a different success path or paths. For a less severe scenario, the success path may encompass all or a portion of the success path(s) for a catastrophic event. Documentation of the success path(s) generally includes a list of systems (i.e. front line and support systems) and an itemization of their functions, designs and dependencies. Two dependency tables can often be created to document the direct dependency of front line systems on support systems, as well as the dependencies among support systems. Additional guidance on documentation is provided in Section 8.

METHODOLOGY FOR COPING WITH VULNERABILITIES OF STRUCTURES, SYSTEMS AND COMPONENTS IMPORTANT TO SAFETY

7.53. Upon completion of the sabotage margin assessment, some vulnerabilities of SSCs important to safety or SSCs with unacceptably low capacities may be identified. Methods should be considered to successfully mitigate the adverse consequences resulting from identified vulnerabilities, taking into account defence in depth for safety and any additional on-site safety measures and security measures that might not have been considered and that might help to mitigate these adverse consequences.

7.54. Available off-site resources should also be identified for mitigating adverse consequences, such as those for emergency response (e.g. fire suppression material, pumps, cables, power supplies, heavy lifting equipment and other equipment that could be used to mitigate the results of damage from a wide range of attack scenarios). Accident management, containment performance and other mitigation measures should be considered in the evaluation process.

7.55. For coping with those vulnerabilities that are identified in relation to SSCs important to safety, the facility management can implement a number of actions, including the following:

(a) Strengthening the physical protection system for vital areas;
(b) Introducing engineered changes to SSCs;
(c) Adjusting the facility to provide a layout that is easier to protect;

(d) Upgrading the facility's accident management and emergency response capabilities;
(e) Building the capacity of the on-site and off-site response.

7.56. Factors that should be taken into consideration by the facility management in selecting the actions for coping with vulnerabilities include the severity of the assessed vulnerabilities, the options available for facility upgrades, and the optimization of the allocation of resources needed for upgrades and changes, on the basis of estimated improvements in terms of the prevention or mitigation of the consequences of a successful sabotage attack. Use of available severe accident management capabilities and strengthening of off-site capabilities are also important factors to consider.

7.57. Options available for facility upgrades may include physical protection upgrades, safety system upgrades (e.g. redundancy, diversity and separation measures) and structural strengthening (e.g. efforts targeting the survival of a given SSC). In addition, if there is no warning time prior to the sabotage attack, operator action during or after the attack may also be essential in diagnosing and responding to the event, if this is not performed automatically.

7.58. Decisions to allocate the resources needed for facility upgrades and changes should be based on the following:

(a) Capability of the improvement to prevent or mitigate sabotage consequences.
(b) Estimated improvements in the performance of the SSCs (e.g. margin improvements).
(c) Ease of implementation of the potential improvement.
(d) Time needed for the completion of the upgrade (e.g. outage).
(e) Duration of time that the type 2 threat scenario is deemed to be credible in relation to the potential risk; for example, short term risks would necessitate short term measures whereas long term risks would necessitate long term measures.

7.59. Mitigating features, such as existing accident management procedures, may need to be enhanced to better address any command and control issues and emergency plan implementation under conditions that may include partial or complete loss of the main control room or alternative shutdown panel functionality, the technical support centre and/or operating personnel. Such enhancements should be complemented by other measures such as alerting firefighters and increasing the number of response force personnel on the site.

7.60. It may also be possible to strengthen off-site capabilities; for example, through the installation of physical barriers, the creation of exclusion zones or the surveillance of access roads to the nuclear facility by law enforcement officers. These measures could ultimately reduce the potential for and the severity of attacks.

7.61. For some type 2 sabotage threat scenarios, protection based on facility resources (e.g. engineering upgrades, enhancements of physical protection) could be problematic, so discussions with competent authorities may also be necessary to determine how to respond to these specific cases. As an alternative, the competent authorities may decide to implement additional off-site prevention or response measures.

8. DOCUMENTATION AND INFORMATION SECURITY

8.1. Thorough documentation of the process undertaken to identify and categorize sabotage targets and to identify vital areas at the facility should be maintained.

8.2. The documentation maintained in relation to this process should contain information on the following:

(a) Identification and evaluation of inventories of nuclear and other radioactive material;
(b) Determination of potential radiological consequences of sabotage attacks;
(c) Consideration of sabotage attack scenarios;
(d) Consideration of operational activities;
(e) Identification of sabotage targets that could lead to unacceptable radiological consequences or high radiological consequences;
(f) Identification of initiating events of malicious origin;
(g) Identification of safety systems that could mitigate initiating events of malicious origin;
(h) Identification of sabotage targets that are needed for preventing high radiological consequences;
(i) Definition of a safe state for the facility;
(j) Identification of sabotage targets needed to maintain the facility in a safe state;
(k) Development of the sabotage logic model;
(l) Facility walkdown and identification of areas;

(m) Identification of the location of the SSCs important to safety included in the safe shutdown equipment list;
(n) Identification of candidate vital area sets;
(o) Selection of a set of vital areas;
(p) Determination of measures for detection, access control and delay for the protection of SSCs important to safety;
(q) Vulnerability evaluation and sabotage margin assessment;
(r) Establishment of the roles and responsibilities of the emergency response organization(s);
(s) Drafting of a list of procedures to manage the interface between nuclear safety and nuclear security.

8.3. The organization of the documentation should be governed by four general principles:

(1) Traceability: Information should be traceable with a minimum of effort in order to provide the user of the documentation with the ability to review and update the analysis.
(2) Sequential order: The order in which the analysis is recorded in the report should follow the order in which the analysis was performed.
(3) Transparency: The applied thresholds and acceptance criteria, dates and names of the individuals involved in the analysis should be documented.
(4) Confidentiality: The information and the documentation developed should be appropriately secured and administrated.

SECURITY OF SENSITIVE INFORMATION

8.4. The process for sabotage target identification and categorization and for vital area identification generates sensitive information that is required to be properly protected according to the information security requirements established by the competent authority. The information security requirements and procedures depend on the legal system in the State where the facility is located. All those who have access to the information generated in the process of identifying sabotage targets and vital areas should understand and follow the relevant information security requirements, such as the 'need to know' rule. Guidance on the security of nuclear information and information assets is provided in IAEA Nuclear Security Series No. 23-G, Security of Nuclear Information [32].

8.5. All the documentation containing physical protection information that is generated through this process — whether on an individual basis or assembled for the purpose of assessing engineering safety aspects — should be considered security sensitive information and thus should be appropriately secured.

8.6. Procedures should be in place to minimize the risk of disclosure of confidential information. Keeping security information separate from facility condition information supports the application of the need to know principle.

8.7. The walkdown team and other personnel supporting the walkdown team (e.g. administrative support) should consist of staff whose trustworthiness has been assessed. Further guidance on assessing trustworthiness is contained in paras 4.13–4.18 of Ref. [27].

REFERENCES

[1] The Convention on the Physical Protection of Nuclear Material, INFCIRC/274/Rev. 1, IAEA, Vienna (1980).

[2] Amendment to the Convention on the Physical Protection of Nuclear Material, INFCIRC/274/Rev.1/Mod. 1 (Corrected), IAEA, Vienna (2021).

[3] INTERNATIONAL ATOMIC ENERGY AGENCY, Nuclear Security Recommendations on Physical Protection of Nuclear Material and Nuclear Facilities (INFCIRC/225/Revision 5), IAEA Nuclear Security Series No. 13, IAEA, Vienna (2011), https://doi.org/10.61092/iaea.ko2c dc4q

[4] INTERNATIONAL ATOMIC ENERGY AGENCY, Physical Protection of Nuclear Material and Nuclear Facilities (Implementation of INFCIRC/225/Revision 5), IAEA Nuclear Security Series No. 27-G, IAEA, Vienna (2018).

[5] INTERNATIONAL ATOMIC ENERGY AGENCY, National Nuclear Security Threat Assessment, Design Basis Threats and Representative Threat Statements, IAEA Nuclear Security Series No. 10-G (Rev. 1), IAEA, Vienna (2021).

[6] INTERNATIONAL ATOMIC ENERGY AGENCY, Handbook on the Design of Physical Protection Systems for Nuclear Material and Nuclear Facilities, IAEA Nuclear Security Series No. 40-T, IAEA, Vienna (2021).

[7] INTERNATIONAL ATOMIC ENERGY AGENCY, Nuclear Security Recommendations on Radioactive Material and Associated Facilities, IAEA Nuclear Security Series No. 14, IAEA, Vienna (2011).

[8] INTERNATIONAL ATOMIC ENERGY AGENCY, Computer Security Techniques for Nuclear Facilities, IAEA Nuclear Security Series No. 17-T (Rev. 1), IAEA, Vienna (2021).

[9] EUROPEAN ATOMIC ENERGY COMMUNITY, FOOD AND AGRICULTURE ORGANIZATION OF THE UNITED NATIONS, INTERNATIONAL ATOMIC ENERGY AGENCY, INTERNATIONAL LABOUR ORGANIZATION, INTERNATIONAL MARITIME ORGANIZATION, OECD NUCLEAR ENERGY AGENCY, PAN AMERICAN HEALTH ORGANIZATION, UNITED NATIONS ENVIRONMENT PROGRAMME, WORLD HEALTH ORGANIZATION, Fundamental Safety Principles, IAEA Safety Standards Series No. SF-1, IAEA, Vienna (2006),
https://doi.org/10.61092/iaea.hmxn-vw0a

[10] INTERNATIONAL ATOMIC ENERGY AGENCY, Objective and Essential Elements of a State's Nuclear Security Regime, IAEA Nuclear Security Series No. 20, IAEA, Vienna (2013),
https://doi.org/10.61092/iaea.ajrj-ymul

[11] INTERNATIONAL ATOMIC ENERGY AGENCY, Developing a Nuclear Security Contingency Plan for Nuclear Facilities, IAEA Nuclear Security Series No. 39-T, IAEA, Vienna (2019).

[12] FOOD AND AGRICULTURE ORGANIZATION OF THE UNITED NATIONS, INTERNATIONAL ATOMIC ENERGY AGENCY, INTERNATIONAL CIVIL AVIATION ORGANIZATION, INTERNATIONAL LABOUR ORGANIZATION, INTERNATIONAL MARITIME ORGANIZATION, INTERPOL, OECD NUCLEAR ENERGY AGENCY, PAN AMERICAN HEALTH ORGANIZATION, PREPARATORY COMMISSION FOR THE COMPREHENSIVE NUCLEAR-TEST-BAN TREATY ORGANIZATION, UNITED NATIONS ENVIRONMENT PROGRAMME, UNITED NATIONS OFFICE FOR THE COORDINATION OF HUMANITARIAN AFFAIRS, WORLD HEALTH ORGANIZATION, WORLD METEOROLOGICAL ORGANIZATION, Preparedness and Response for a Nuclear or Radiological Emergency, IAEA Safety Standards Series No. GSR Part 7, IAEA, Vienna (2015),
https://doi.org/10.61092/iaea.3dbe-055p

[13] INTERNATIONAL ATOMIC ENERGY AGENCY, Arrangements for Preparedness for a Nuclear or Radiological Emergency, IAEA Safety Standards Series No. GS-G-2.1, IAEA, Vienna (2007).

[14] INTERNATIONAL ATOMIC ENERGY AGENCY, Considerations in the Development of a Protection Strategy for a Nuclear or Radiological Emergency, IAEA EPR-Protection Strategy 2020, IAEA, Vienna (2021).

[15] INTERNATIONAL ATOMIC ENERGY AGENCY, Safety of Conversion Facilities and Uranium Enrichment Facilities, IAEA Safety Standards Series No. SSG-5 (Rev. 1), IAEA, Vienna (2023).

[16] INTERNATIONAL ATOMIC ENERGY AGENCY, Safety of Uranium Fuel Fabrication Facilities, IAEA Safety Standards Series No. SSG-6 (Rev. 1), IAEA, Vienna (2023).

[17] INTERNATIONAL ATOMIC ENERGY AGENCY, Storage of Spent Nuclear Fuel, IAEA Safety Standards Series No. SSG-15 (Rev. 1), IAEA, Vienna (2020).

[18] INTERNATIONAL ATOMIC ENERGY AGENCY, Use of a Graded Approach in the Application of the Safety Requirements for Research Reactors, IAEA Safety Standards Series No. SSG-22 (Rev. 1), IAEA, Vienna (2023).

[19] INTERNATIONAL ATOMIC ENERGY AGENCY, Safety of Nuclear Fuel Reprocessing Facilities, IAEA Safety Standards Series No. SSG-42 (Rev. 1), IAEA, Vienna (in preparation).
[20] INTERNATIONAL ATOMIC ENERGY AGENCY, Safety of Nuclear Power Plants: Design, IAEA Safety Standards Series No. SSR-2/1 (Rev. 1), IAEA, Vienna (2016).
[21] INTERNATIONAL ATOMIC ENERGY AGENCY, Safety of Research Reactors, IAEA Safety Standards Series No. SSR-3, IAEA, Vienna (2016).
[22] INTERNATIONAL ATOMIC ENERGY AGENCY, Safety of Nuclear Fuel Cycle Facilities, IAEA Safety Standards Series No. SSR-4, IAEA, Vienna (2017).
[23] INTERNATIONAL ATOMIC ENERGY AGENCY, IAEA Nuclear Safety and Security Glossary: Terminology Used in Nuclear Safety, Nuclear Security, Radiation Protection and Emergency Preparedness and Response, 2022 (Interim) Edition, IAEA, Vienna (2022),
https://doi.org/10.61092/iaea.rrxi-t56z
[24] INTERNATIONAL ATOMIC ENERGY AGENCY, Format and Content of the Safety Analysis Report for Nuclear Power Plants, IAEA Safety Standards Series No. SSG-61, IAEA, Vienna (2021).
[25] INTERNATIONAL ATOMIC ENERGY AGENCY, Safety Assessment for Facilities and Activities, IAEA Safety Standards Series No. GSR Part 4 (Rev. 1), IAEA, Vienna (2016).
[26] INTERNATIONAL ATOMIC ENERGY AGENCY, Design of Nuclear Installations Against External Events Excluding Earthquakes, IAEA Safety Standards Series No. SSG-68, IAEA, Vienna (2021).
[27] INTERNATIONAL ATOMIC ENERGY AGENCY, Preventive and Protective Measures against Insider Threats, IAEA Nuclear Security Series No. 8-G (Rev. 1), IAEA, Vienna (2020).
[28] INTERNATIONAL ATOMIC ENERGY AGENCY, Developing a National Framework for Managing the Response to Nuclear Security Events, IAEA Nuclear Security Series No. 37-G, IAEA, Vienna (2019).
[29] INTERNATIONAL ATOMIC ENERGY AGENCY, Safety Aspects of Nuclear Power Plants in Human Induced External Events: Assessment of Structures, Safety Reports Series No. 87, IAEA, Vienna (2018).
[30] INTERNATIONAL ATOMIC ENERGY AGENCY, Seismic Hazards in Site Evaluation for Nuclear Installations, IAEA Safety Standards Series No. SSG-9 (Rev. 1), IAEA, Vienna (2022).
[31] INTERNATIONAL ATOMIC ENERGY AGENCY, Protection against Internal Hazards in the Design of Nuclear Power Plants, IAEA Safety Standards Series No. SSG-64, IAEA, Vienna (2021).
[32] INTERNATIONAL ATOMIC ENERGY AGENCY, Security of Nuclear Information, IAEA Nuclear Security Series No. 23-G, IAEA, Vienna (2015).

Annex I

EXAMPLE OF A SABOTAGE LOGIC MODEL

I–1. A step by step solution for a simple sabotage logic model is provided in this annex to illustrate how candidate vital area sets can be identified. The example logic model demonstrates how the concepts of minimum cutsets and minimum protection sets are applied in the vital area identification process. A logic model can be a statement, an algebraic expression or a graphical representation, such as a fault tree or an event tree. The solution for different representations of the same logic problem will give the same results.

I–2. A logic model is 'solved' by applying the rules of Boolean algebra to the model. Table I–1 provides definitions of common logic symbols and Boolean algebra rules. Consider a fictitious facility that has the following characteristics:

(a) Two initiating events are identified for this facility: initiating event 1 (IE1) and initiating event 2 (IE2). If unmitigated, these events will result in releases that exceed the threshold for high radiological consequences established by the competent authority.

(b) Safety system 1 (S1) is designed to mitigate IE1 and safety system 2 (S2) is designed to mitigate IE2.

(c) S1 has two trains of equipment, train 1 (T1) and train 2 (T2). If either of these trains functions properly, S1 can successfully mitigate IE1 (i.e. both trains have to fail for S1 to fail).

(d) S2 has three trains: T3, T4 and T5. Either T3 or both T4 and T5 have to function in order for S2 to successfully mitigate IE2 (i.e. S2 will fail to mitigate IE2 if either T3 and T4 fail or if T3 and T5 fail).

(e) The trains in the systems have components (C) that need to operate for the trains to function.
    (i) T1 fails if either of the two components C1 or C2 fails;
    (ii) T2 fails if either C3 or C4 fails;
    (iii) T3 fails if either C5 or C6 fails;
    (iv) T4 fails if either C7 or C8 fails;
    (v) T5 fails if either C9 or C10 fails.

(f) In order for the adversary to cause the initiating event and disable the various components, the person(s) in question would have to gain access to different facility locations (L) as shown in Table I–1.

TABLE I–1. EXAMPLE SABOTAGE SCENARIO

Action	Location
Disable C1	L1
Disable C2	L2
Disable C3	L2
Disable C4	L2
Disable C5	L3
Disable C6	L3
Disable C7	L5
Disable C8	L6
Disable C9	L6
Disable C10	L6
Cause IE1	L8
Cause IE2	L9

I–3. The statements in para. I–2 constitute one form of logic model for the sabotage of a facility. By carefully analysing these statements, the combinations of locations that an adversary would have to enter to cause all of the initiating events and component failures that would lead to high radiological consequences can be determined. For example, if an adversary could gain access to L2 and L8, the adversary could initiate IE1 and disable S1, resulting in a release that exceeds high radiological consequence limits. The adversary can cause IE1 if access is gained to L8. If the adversary disables both T1 and T2, S1 will not be able to mitigate IE1. Disabling C2 can disable T1, and disabling C3 can disable T2. Both C2 and C3 can be disabled from L2, so by gaining access to both L2 and L8, the adversary could cause high radiological consequences. By reviewing the statements, actions and locations in detail, all of the combinations of locations from which initiating events could occur and are sufficient to cause high radiological consequences can be identified. As long as the facility is simple enough, it is possible to derive by inspection the location combinations in which sabotage could be accomplished, as described above. A more useful approach would be to represent the relationships among initiating events, disablement actions and locations in a logic equation. The event to be represented in this logic equation is a release in excess of high radiological consequences. Using the definitions provided in Table I–1, the following equations are developed:

$$HRC = IE1 * S1 + IE2 * S2 \quad (I–1)$$

where HRC denotes high radiological consequences.

$$S1 = T1 * T2 \quad (I–2)$$

$$S2 = T3 * T4 + T3 * T5 \quad (I–3)$$

$$T1 = C1 + C2 \quad (I–4)$$

$$T2 = C3 + C4 \quad (I–5)$$

$$T3 = C5 + C6 \quad (I–6)$$

$$T4 = C7 + C8 \quad (I–7)$$

$$T5 = C9 + C10 \quad (I–8)$$

I–4. Equations (I–1) to (I–8) indicate that S1 is disabled, T1 is disabled and C1 is disabled. Replacing the events in these equations with the locations in which they can be caused and simplifying by using the rules of Boolean algebra yields the following results:

$$T1 = L1 + L2 \quad (I–9)$$

$$T2 = L2 + L2 = L2 \quad (I–10)$$

$$T3 = L3 + L3 = L3 \quad (I–11)$$

$$T4 = L5 + L6 \quad (I–12)$$

$$T5 = L6 + L6 = L6 \quad (I–13)$$

$$S1 = (L1 + L2) * L2 = L2 \quad (I–14)$$

$$S2 = L3 * (L5 + L6) + L3 * L6 = L3 * L5 + L3 * L6 \quad (I–15)$$

$$\text{HRC} = L8 * L2 + L9 * (L3 * L5 + L3 * L6) \qquad \text{(I–16)}$$

$$= (L8 * L2) + (L9 * L3 * L5) + (L9 * L3 * L6)$$

I–5. For this simple example, there are three combinations of locations in which an adversary could cause high radiological consequences:

$$\text{HRC} = L8 * L2 + L9 * L3 * L5 + L9 * L3 * L6 \qquad \text{(I–17)}$$

I–6. Each combination of locations in which sabotage can be caused is called a cutset of the sabotage location equation. The objective of vital area identification is to find a minimum set of areas in which targets are to be protected against sabotage to prevent all of the possible sabotage attack scenarios that could lead to high radiological consequences. In other words, at least one of the areas in each combination of areas in which sabotage can be accomplished is determined to be a vital area. Each combination of locations in which protection could prevent all of the sabotage attack scenarios represents a prevention set for the logic model and constitutes a candidate vital area set. For simple sabotage location equations, it is possible to directly determine the combinations of locations in which protection will prevent sabotage. Equation (I–17) demonstrates that if the adversary is prevented from gaining access to the combinations of areas shown in Eq. (I–18), high radiological consequences can be prevented from occurring.

$$\text{HRC prevented}$$

$$= \underline{L8} * \underline{L9} + \underline{L8} * \underline{L3} + \underline{L2} * \underline{L9} + \underline{L2} * \underline{L3} + \underline{L8} * \underline{L5} * \underline{L6} + \underline{L8} * \underline{L5} * \underline{L6} \qquad \text{(I–18)}$$

I–7. In Eq. (I–18), the underlining indicates that access to the location is prevented; for example, $\underline{L8}$ means that access to L8 is prevented. In Boolean algebra terms, $\underline{L8}$ is the complement (non-occurrence or NOT) of L8. For the example facility, there are six candidate vital area sets, as shown in Eq. (I–18). This result can also be derived algebraically by forming the complement of the sabotage location equation and by simplifying through use of the rules of Boolean algebra. The protection of any one of the candidate vital area sets ensures that an adversary cannot cause high radiological consequences. If, for example, the set of L2 and L3 is selected as the final vital area set, these are the only two areas of the facility that will be protected as vital areas. Protecting these two areas will ensure that none of the possible sabotage attack scenarios can be completed.

TABLE I–2. ATTACK PATHWAY KEY

	Logic symbols	
Symbol	Operation	Definition
+	OR	Either of two events occurs. A + B means that either event A or event B occurs.
*	AND	Both of two events occur. A * B means that both event A and event B occur.
	Logic gates	
Symbol	Gate name	Definition
	OR gate	Output occurs if any of the inputs occur.
	AND gate	Output occurs if all of the inputs occur.
	Boolean algebra rules	
A + A = A	A +A * B = A	$(\underline{A + B}) = \underline{A} * \underline{B}$
A * A = A	A * (B + C) = A * B + A * C	$(\underline{A * B}) = \underline{A} + \underline{B}$

I–8. Fault trees can be used to efficiently represent the sabotage logic for more complicated facilities. Figure I–1 provides a fault tree for the example facility that will be solved to further illustrate the process of identifying candidate vital area sets. The top event in this tree is a release in excess of high radiological consequence limits. The logic gates show the ways in which the events in the tree combine to cause the top event, and the tree is developed down to the level of component failures. Figure I–2 shows the fault tree with all the terminal events replaced with the locations in which the events can be caused. This sabotage location fault tree is solved using the Boolean algebra concepts applied in Eqs (I–1) to (I–17) to produce the same results. The expression in parentheses beside each gate is the solution for the gate in terms of the terminal events in the tree. One way to generate the level 1 protection sets for a fault tree is to form and solve the dual of the tree. The dual of a fault tree is formed by changing each OR gate in the tree to an AND gate, each AND gate to an OR gate and each event to the complement (NOT) of the event. A variety of software packages are available for solving fault trees and generating the prevention sets (candidate vital area sets) needed in the vital area identification process. In summary, the sabotage logic model for a facility can be developed in several equivalent forms. The solution of the logic model produces candidate vital area sets that can be protected to prevent sabotage. Any one of the candidate sets will contain the minimum set of equipment needed to ensure that no sabotage attack scenarios can be completed.

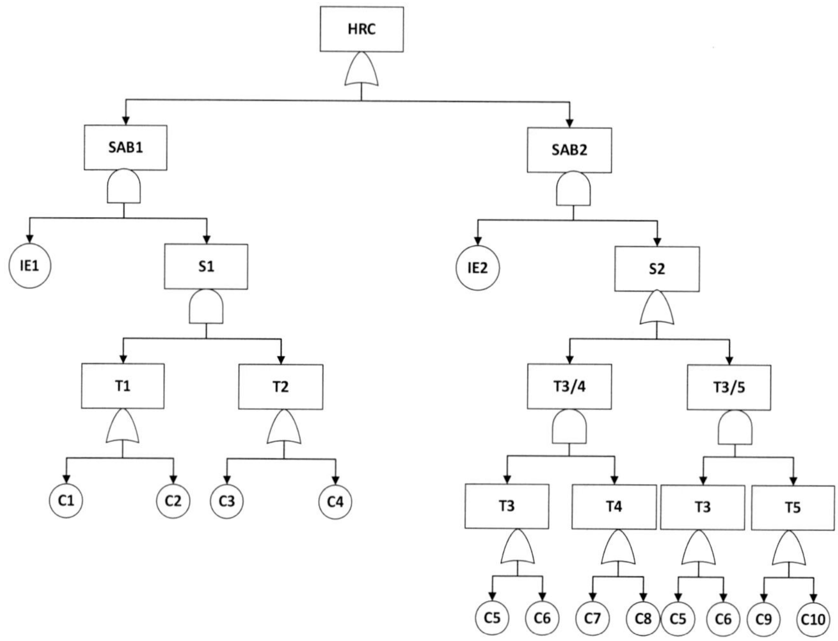

FIG. I–1. *Attack pathway for high radiological consequences; example 1.*

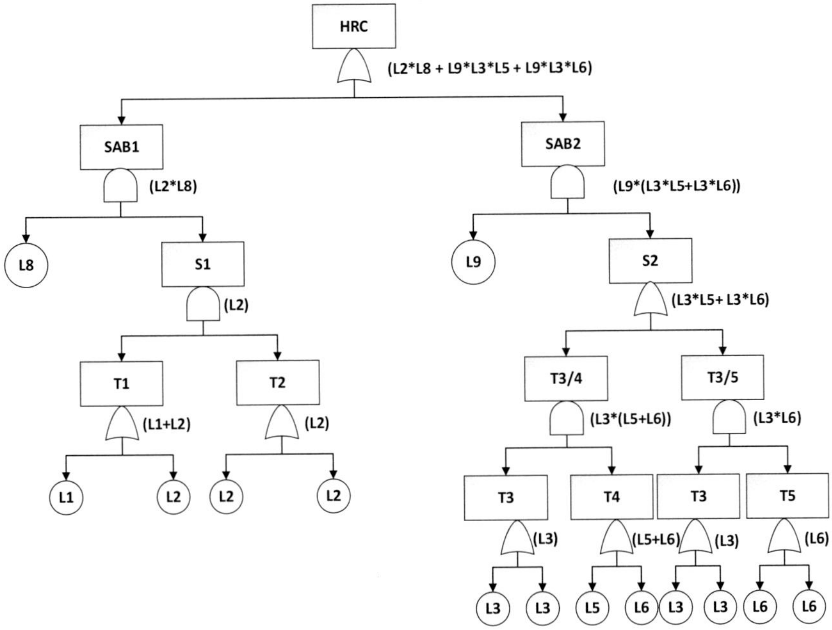

FIG. I–2. Attack pathway for high radiological consequences; example 2.

Annex II

EXAMPLE OF A FACILITY WALKDOWN

FACILITY WALKDOWN TEAMS

II–1. Facility walkdown teams consist of members of the facility operating personnel and consultants with specific expertise relating to topics such as sabotage protection, extreme environmental loads, design and operation of structures, systems and components (SSCs) and safe shutdown conditions. The tasks and responsibilities of the team members are as follows:

(a) Team leader: The team leader supervises the field activities, engineering evaluations and fulfilment of security requirements. Because of the sensitive nature of this effort, the activities associated with walkdowns need to be performed in a focused and secure manner to ensure control over all of the related information. The team leader has to be trustworthy — preferably an employee of the operator — with the authority, supervisory skills and appropriate engineering background to lead walkdowns, as well as with a thorough understanding of the security information control necessary to supervise activities and ensure the security and integrity of the process. The team leader may interact with competent authorities, as necessary, to define or clarify the elements of the sabotage attack scenarios to be evaluated.
(b) Engineering safety experts: Engineering safety experts comprise the walkdown team, which is focused on engineering safety aspects. They may belong to the operating personnel or, if necessary, they may be consultants with expertise in engineering safety. The engineering disciplines to be represented are systems, civil, structural, mechanical, electrical, and instrumentation and control engineering. Radiation protection experts are also needed. All these disciplines are considered in each evaluation to ensure completeness. All engineering safety experts and radiation protection experts need to undergo a trustworthiness assessment (see IAEA Nuclear Security Series No. 8-G (Rev. 1), Preventive and Protective Measures Against Insider Threats [II–1]), which is performed by the operator or other relevant organization (e.g. the regulatory body) and need to obtain the proper clearance and training to maintain the security and integrity of the process.
(c) Members of operating personnel: Members of operating personnel are an essential component of the team, and they can provide their expertise throughout the facility walkdown activities.

II–2. When the physical protection system is evaluated in conjunction with the engineering safety aspects, an integrated team, which includes physical protection experts, may be formed for conducting facility walkdowns. Forming such a team is particularly desirable if the sabotage attack scenarios include multimode attacks that encompass combined sabotage attack scenarios.

II–3. Team members, including the team leader, are to be assigned to the walkdown effort for as long as their involvement is needed, with minimal collateral duties.

FACILITY WALKDOWN PROCEDURE

II–4. The facility walkdown procedure comprises the walkdown preparation, the preliminary screening walkdown and the detailed screening walkdown (see paras II–5 to II–18). Facility walkdown activities and controls benefit from a separate secure workplace that ensures the security and integrity of the effort and related documentation.

Walkdown preparation

II–5. Walkdown preparations include familiarization of the facility walkdown team with the facility, consisting of the following activities:

(a) Assembling general facility documentation, which may include safety analysis reports, system descriptions, piping and instrumentation diagrams, electrical single line diagrams, operating procedures, drawings of the general arrangement of the facility, drawings of mechanical and electrical equipment locations, probabilistic safety assessments for internal and external events, and any other documentation of beyond safety design basis assessments.
(b) Assembling physical protection system information, in particular in relation to the designated vital area set to prevent initiating events of malicious origin.
(c) Determining that the facility requirements have been met by the team members, including in relation to radiation protection (e.g. adherence to the ‘as low as reasonably achievable’ principle), safety practices and security practices.

II–6. The team members are to consult or create facility documents on success paths and the SSCs in the safe shutdown equipment list (SSEL). The environmental load on each SSC in the list is also defined.

II–7. A database of the SSEL is to be prepared, summarizing the evaluation of each SSC in the list in terms of the environmental loads. It is expected that the SSEL of a nuclear power plant will comprise a few hundred SSCs. Other facility types may have significantly fewer SSCs in their lists.

II–8. Individual SSC data sheets are to be prepared, containing information such as the type of the SSC, its function, its location and its potential vulnerabilities. If necessary, the data can be supplemented with field and office generated SSC specific evaluations, including field notes; safety, security and engineering analyses; and field modifications.

II–9. A facility walkdown plan is to be developed to indicate the number of teams and the composition of each team. It is expected that more than one team will be used, with the total number depending on the issues to be considered, the number and type of experts needed and confidentiality requirements.

II–10. Figure II–1 provides an example of the format that can be used for the SSEL. The columns in Fig. II–1 can be defined as follows:

(a) SSEL No.: The sequential number of the SSEL items in the table.
(b) SSC name: Information describing the SSC (e.g. auxiliary building, diesel generator).
(c) SSC ID No.: A facility specific identifier.
(d) Description: A brief description of the SSC.
(e) Sabotage attack scenario No.: An identifier that is linked to the list of attack scenarios to be considered.
(f) Location: A series of location identifiers to aid in planning the facility walkdown and evaluating the consequences of a potential attack. It may include vital area identification for physical protection system evaluation.
(g) Physical loading condition: Guidance on the selection of the type of expert needed, on the facility walkdown access and on the combined load conditions to be evaluated (e.g. impact in combination with fire). The subcolumns are described as follows:
    (i) Impact: Direct and indirect impact effects to be considered in the evaluation. Direct impact effects are conditions that may include direct missile impacts. Indirect impact effects are conditions that may include scabbing of concrete and vibration induced loading.
    (ii) Explosion/blast: Explosion or blast effects can be direct or indirect. Direct impact effects are blast pressures, and indirect blast effects are conditions that may include vibration induced loading.
    (iii) Heat/fire: The effects of heat from a fire or of a direct flame on the SSC.

SSEL[a] No.	SSC name	SSC ID No.	Description	Sabotage attack scenario No.	Location				Physical loading condition				
					Building	Elevation	Room/ compartment	Vital area	Impact	Explosion/ blast	Heat/ fire	Smothering	Flooding

[a] SSEL: safe shutdown equipment list.

FIG. II–1. Example of a composite safe shutdown equipment list.

(iv) Smothering: Smothering and related conditions might arise as a result of smoke, toxic chemicals or firefighting techniques. This failure mode might affect personnel or systems; for example, smothering of a diesel generator system could occur if the air intake system became inundated. Control room habitability and on-site security personnel safety is thus to be evaluated.

(v) Flooding: Flooding can result from internal or external sources of water, which may need to be evaluated.

II–11. Figure II–2 provides an example of an individual data sheet that can be used in the evaluation of SSCs in relation to environmental load conditions. In the walkdown preparation stage, the team notes basic information that identifies the SSC under consideration and completes the remainder of the table after the walkdown and evaluations. The data to be collected and evaluated may need to be modified to take into account non-vibrational modes of failure (i.e. environmental conditions, such as heat, humidity and direct impacts).

SSC name: ____________________ SSC ID No.: ______________

SSC description: ____

Location: Bldg ________ Elev ________ Room/compartment/row/col. ____

Threat scenario No./description: ____________________________

Vital area identification: ________

Performance requirements: _________________________________

SUMMARY (capacity versus demand)

Impact loads:

Direct: ______________________________________

Indirect: ____________________________________

Blast loads:

Direct: ______________________________________

Indirect: ____________________________________

Heat/fire loads:

Heat: ____

Fire: ____

FIG. II–2. Example of a screening evaluation worksheet for environmental load conditions.

Preliminary screening walkdown

II–12. The preliminary screening walkdown achieves the following objectives:

(a) Determines the location and accessibility of each SSC on the SSEL for the facility;
(b) Identifies other SSCs that may be needed for safe shutdown, which are to be added to the SSEL;
(c) Reviews and validates the screening of SSCs with respect to capacity considerations;
(d) Identifies potential easy fixes to SSCs;
(e) Groups all of the components belonging to the same system;
(f) Groups components within the same location, particularly in the same vital area, for evaluation of spatially common environments;
(g) Evaluates whether SSC capacity is adequate for the specified threat(s);
(h) Documents conclusions.

II–13. The preliminary screening walkdown visually examines SSCs that are accessible. There are three alternative dispositions for each SSC on the SSEL:

(1) Disposition category 1: For SSCs in this category, capacity is clearly lower than the demand and a modification is needed.
(2) Disposition category 2: The capacity of SSCs in this category is uncertain, and further evaluation is needed to determine whether a modification is needed.
(3) Disposition category 3: For SSCs in this category, the capacity is clearly greater than the demand and the SSC is adequate for the specified threat.

II–14. The preliminary screening walkdown is to be properly documented. The main result of the preliminary walkdown is the identification of SSCs on the SSEL that are clearly robust. These SSCs are categorized as disposition category 3 and are therefore excluded from further evaluation. SSCs in disposition categories 1 and 2 need a more detailed in-office and in-facility evaluation.

Detailed screening walkdown

II–15. The detailed screening walkdown is to be performed for all SSCs whose capacity for the defined environmental loads has not been verified. This includes in-facility evaluations and, in many cases, further analytical calculations and evaluations. Two categories of SSCs result from these evaluations:

(1) SSCs in the first category are those that were not excluded from further consideration during the preliminary walkdown. At this stage, walkdown engineers evaluate these systems and components in more detail and make a judgement as to whether or not the component needs to be further analysed or modified.
(2) For SSCs in the second category, facility modifications are clearly warranted. In these cases, the walkdown engineers suggest that modifications be implemented.

II–16. The detailed screening walkdown has to be properly documented. It is advisable to supplement the documentation with photographic and/or video records. Figure II–2 provides an acceptable form of summary documentation for the entire SSEL. The SSC evaluations may be documented using the form given in Fig. II–2, with supporting material attached.

II–17. Confidentiality of the documentation is to be strictly maintained, with distribution on a need to know basis only.

SPECIAL TOPICS CONSIDERED IN FACILITY WALKDOWNS

Spatial interactions

II–18. The facility walkdown is a key tool for identifying spatial interactions that could potentially affect the performance of the SSEL. The identification and assessment of potential interactions demands good judgement from the walkdown team.

Falling

II–19. Falling is the structural integrity failure of a non-safety or safety related SSC that could hit and damage a safety related SSC. For the interaction to be a danger to an SSC on the SSEL, the impact has to contain considerable energy and the target has to be vulnerable. For example, a light fixture falling on a 10 cm diameter pipe might not be a credible damage threat to the pipe. However, the same light fixture falling on an open relay panel is an interaction that could cause damage and has to be addressed. Scabbing of concrete resulting from missile impact on a building element (e.g. wall, diaphragm, roof) might be a viable failure mode for delicate equipment in the range of the falling concrete. Unreinforced masonry walls are a common source of falling interaction. Masonry walls are generally located close enough to the safety related equipment whose failure could lead to equipment damage.

Proximity

II–20. Proximity interactions are defined as conditions in which two or more SSCs are close enough so that the behaviour of one might have consequences on the behaviour of the others. The most common example of proximity interaction is fire or explosion; more details can be found in IAEA Nuclear Security Series No. 13, Nuclear Security Recommendations on Physical Protection of Nuclear Material and Nuclear Facilities (INFCIRC/225/Revision 5) [II–2].

Spray and flood

II–21. Spray and flood can result from the failure of piping, systems or vessels that are not properly supported or anchored. Inadvertent spray hazards to SSCs on the SSEL are most often associated with wet fire protection piping systems. The most common source of spray is leakage caused by the impact induced failures of sprinkler heads. Since fire and heat are potential safety threats throughout the facility site, particularly in buildings and compartments, the walkdown evaluates the vulnerability to spray for all of the SSEL components. Generally, design evaluations of fire and fire suppression systems will have taken spray vulnerabilities into account. If spray sources can reach equipment sensitive to water spray, then the source is backfitted, usually by adding support to reduce deflections, impact or stress. An alternative is to protect the target — in this case, the SSC.

II–22. Large tanks are potential flood sources. The walkdown team, with the assistance of facility personnel, assesses the potential consequences of a flood source failure and the ability of the floor drainage system to mitigate the consequences of a source failure.

Type and number of co-located facilities at the site

II–23. A nuclear power plant site may have several reactor units, possibly with interdependent safety or support systems; multiple unit nuclear power plant sites often assume the availability of companion unit systems when addressing events that are not considered common cause failures. In addition, other critical facilities may be present within the site boundary, such as spent fuel storage in fuel pools or dry cask storage. Research reactor sites may have associated laboratories, isotope production facilities and hot cells. The evaluation takes into consideration all of the on-site facilities, including any interdependence of their safety systems. Such considerations include consequence evaluation of environmental discharges that are cumulative for all the facilities at the site.

REFERENCES TO ANNEX II

[II–1] INTERNATIONAL ATOMIC ENERGY AGENCY, Preventive and Protective Measures Against Insider Threats, IAEA Nuclear Security Series No. 8-G (Rev. 1), IAEA, Vienna (2020).

[II–2] INTERNATIONAL ATOMIC ENERGY AGENCY, Nuclear Security Recommendations on Physical Protection of Nuclear Material and Nuclear Facilities (INFCIRC/225/ Revision 5), IAEA Nuclear Security Series No. 13, IAEA, Vienna (2011), https://doi.org/10.61092/iaea.ko2c-dc4q

Annex III

EXAMPLE OF AN EXTREME ENVIRONMENTAL LOAD EVALUATION

III–1. Figure III–1 contains an example matrix for the evaluation of the load resulting from an aircraft crash; recommendations for the assessment of the load (e.g. including structures) are provided in IAEA Safety Standards Series No. SSG-9 (Rev. 1), Seismic Hazards in Site Evaluation for Nuclear Installations [III–1]. Depending on the State's threat assessment and sabotage attack scenarios to be considered, such accident events could be used to evaluate attack scenarios resulting in extreme environmental loads.

III–2. The columns in Fig. III–1 can be defined as follows:

(a) Sabotage attack scenario No.: A numerical identifier with values ranging between 1 and the total number of scenarios considered. In the example, sabotage attack scenario number 1 is assumed.
(b) Sabotage attack scenario description: A brief depiction of the sabotage attack scenario for identification purposes. In the example, the scenario involves the impact of a fully fuelled aircraft flown into the nuclear power plant site.
(c) Environmental load condition: Numerical identifiers of the type and specifics of load conditions imposed by the sabotage attack scenario. The identifiers correlate directly with Fig. III–2 for impact, Fig. III–3 for explosion/blast, Fig. III–4 for heat/fire, Fig. III–5 for hazardous material release and Fig. III–6 for other environmental consequences. Figure III–6 provides an example of the engineering disciplines needed for the evaluation, as well as background information regarding why certain environmental load combinations need to be considered. The subcolumns are described as follows:
    (i) Impact: The impact load condition(s), identified by number and by reference to Fig. III–2. In the example, impact load conditions 1 and 2 are assumed.
    (ii) Explosion/blast: The explosion/blast load condition(s), which are identified by number and by reference to Fig. III–3. In the example, no blast or explosion loads are associated with sabotage attack scenario No. 1 and no ancillary effects from the aircraft impact are considered.
    (iii) Heat/fire: The heat and fire load condition(s), identified by number and by reference to Fig. III–4. In the example, heat and fire environmental load condition 1 is assumed.

Sabotage attack scenario No.	Sabotage attack scenario description	Environmental load condition						
		Impact (Fig. III–2)	Explosion/blast (Fig. III–3)	Heat/fire (Fig. III–4)	Hazardous material release (Fig. III–5)	Smothering (Fig. III–6)	Flooding (Fig. III–6)	Other (Fig. III–6)
1	Impact of fully fuelled aircraft flown into nuclear power plant site	1, 2	None	1	None	None	None	None
2	Shoulder launched missile fired into reactor building							
3	Truck explosion at site gate							

FIG. III–1. Example of an extreme environmental load matrix.

(iv) Hazardous material release: The hazardous material release condition(s), identified by number and by reference to Fig. III–5. In the example, no hazardous material release condition is associated with sabotage attack scenario No. 1.

(v) Smothering, flooding and other phenomena: These are identified in Fig. III–1 as examples for future consideration. Smothering, choking and depriving structures, systems and components (SSCs) of the air necessary for operation are suggested as potential concerns; for example, lack of air to diesel generators could prevent startup and operation. Smothering resulting from firefighting techniques (foam) may need to be evaluated. Flooding of the site from internal or external sources may also need to be evaluated; for example, the sabotage of an upstream dam could release a large quantity of water to flood the site.

III–3. Figure III–2 identifies the impact parameters to be used by facility engineers for the evaluation of SSCs. The sabotage attack scenario example from Fig. III–1 is continued in Fig. III–2 for illustrative purposes only.

III–4. The columns of Fig. III–2 can be defined as follows:

(a) Missile type/No.: The missile load identifier, with values ranging between 1 and the total number of missile impact scenarios considered.
(b) Description: A brief description of the source of the load condition.
(c) Mass/weight: The mass or weight of the missile.
(d) Missile characteristics: This includes the following four entries:
   (i) Shape/configuration: A more specific description of the missile, including the dimensions specified, if available. In the example, missile No. 1 is described as having a flexible fuselage, with dimensions to be determined; for missile No. 2, the engines are assumed to be rigid, with dimensions as shown.
   (ii) Impact angle: The angle or range of potential impact angles, taking into account the physics and human capability necessary to achieve the objective.
   (iii) Impact velocity: The velocity of the missile, taking into account the physics and human capability necessary to achieve the objective.
   (iv) Relative hardness: An important parameter for assessing the effect of the missile on SSCs; it can be a qualitative or quantitative measure.
(e) Missile effects: Three entries are included under this heading. These are consequences of direct impact — such as spalling or scabbing of concrete — and have an ancillary effect on components in the neighbourhood

Missile type/No.	Description	Mass/weight	Missile characteristics				Missile effect		
			Shape/ configuration	Impact angle	Impact velocity	Relative hardness	Fire	Explosion/ blast	Other
1	Aircraft fuselage, fully fuelled	200 000 kg	Flexible	Less than 30° from horizontal	180 m/s	Flexible	1	None	None
2	Aircraft engines as projectiles	3 500 kg	3 m diameter rigid cylinder	Less than 30° from horizontal	180 m/s	Rigid	None	None	None
3									

FIG. III–2. Example of an impact parameter definition matrix.

of the impact. They may be specified in other places in the specification; fire is the example used here.

(i) Fire: The missile impact causes a fire either by carrying a combustible or by impacting a combustible, such as a diesel oil tank. Missile No. 1 in the example is associated with heat/fire condition 1, which is a jet fuel fire resulting from an aircraft impact. Missile No. 2 has no related fire condition.

(ii) Explosion/blast: The missile impact causes an explosion or blast, either because the missile is carrying explosives, which detonate upon impact, or because the missile impacts an explosives storage facility.

(iii) Other: Other hazards can include adversaries working in coordination with those responsible for the missile attack.

III–5. Figure III–3 identifies a simplified set of parameters for explosion/blast load conditions to be used by facility engineering for the evaluation of SSC capacity. In the example used here, no explosion/blast conditions were assumed.

The columns in Fig. III–3 can be defined as follows:

(a) Explosion No.: The explosion/blast condition identifier for the blast conditions considered in the example.

(b) The parameters in Fig. III–3 are examples of descriptors of the explosives' characteristics. For general descriptions, TNT equivalent and reference distance (measured from a facility reference point) can be considered as general information (columns 2–4). Specific information about the incident and reflected waves, provided under the heading 'Pressure pulse', would be developed for individual nuclear power plants under evaluation. The details are a function of numerous site specific characteristics.

III–6. Figure III–4 identifies the heat and fire characteristics to be used by facility engineers for the evaluation of SSCs.

Explosion No.	Description	TNT equivalent	Reference distance	Pressure pulse	
				Incident	Reflected
1					
2					
3					

FIG. III–3. Example of an explosion/blast parameter definition matrix.

Fire No.	Description	Fire source outside facility						Fire source or combustibles inside facility				
		Combustible/ ignition	Type	Quantity	Heat potential/ temperature	Duration of burn	Other	Building/ yard	Quantity	Type	Ignition likelihood	Duration of burn
1	Jet fuel fire from aircraft	Yes	Class B (flammable liquids)	50 000 kg	1000°C	1–8 h						
2												
3												

FIG. III–4. Example of a heat/fire parameter definition matrix.

III–7. The columns in Fig. III–4 can be defined as follows:

(a) Fire No.: The heat/fire condition identifier, with values ranging between 1 and the total number of fire conditions considered.
(b) Description: A brief description of the source of the fire.
(c) Fire source outside facility: These entries define the fire hazard, assuming that the source is outside the facility. For an aircraft impact or other similar sabotage attack scenario, the distribution of combustibles within and outside the facility boundary is important. Two obvious distributions are that in the facility yard and the penetration into buildings. Others include those distributions outside the facility boundaries that could inhibit access by emergency responders and others. Examples of important parameters that can be used as subheadings are the quantity and type of combustible or ignition, estimates of the heat potential and temperature, and the duration of burn. The example considers that jet fuel from an aircraft is spilled and ignited, and there is no penetration into buildings. The quantity of fuel is 50 000 kg. The duration of burn at high temperature (1000°C) is 1 hour maximum, with 5–7 hours of residual fire at 300°C.
(d) Fire source or combustibles inside facility: These entries define the fire hazards, assuming that the source is inside the facility or that the fire is ignited inside the facility as a consequence of an outside source. Examples of important parameters are the type and quantity of combustible, the location (e.g. building, yard) and the estimated duration of the burn.

III–8. Figure III–5 identifies important parameters for hazardous material release conditions at a nuclear power plant. Hazardous material releases in conjunction with other modes of simultaneous attack appear to be credible; other modes could include adversaries protected against the effects of the chemical releases. No hazardous material release was assumed in the example presented in Fig. III–5 and para. III–9.

III–9. The columns in Fig. III–5 can be defined as follows:

(a) Case No.: The hazardous material release number, with values ranging between 1 and the total number of hazardous material release conditions considered.
(b) Material description: A brief description of the hazardous material.
(c) Hazardous material load conditions: These include the following:
    (i) Quantity: The amount of the material released and the time frame over which the release occurs;

Case No.	Material description	Hazardous material load conditions						
		Quantity	Smothering effect — personnel	Smothering effect — components	Lethal or disabling effect — personnel	Duration	Extent of penetration	Other
1								
2								
3								
4								

FIG. III–5. Example of a hazardous material release definition matrix.

Plant area		Vital area	Description	Environmental load condition						
				Impact	Explosion/ blast	Heat/fire	Hazardous material release	Smothering	Flooding	Other
Building	1									
	2									
	3									
Zone	1									
	2									
	3									
	4									
Yard	1									
	2									
SSEL[a] item	1									
	2									
	3									

[a] SSEL: safe shutdown equipment list.

FIG. III–6. Example of an extreme environmental load definition matrix.

(ii) Smothering effect — personnel: An itemization of the physical effects on personnel (e.g. facility operating personnel, security forces), including an indication of whether protective gear is needed and the time frame of implementation;
(iii) Smothering effect — components: The potential effects of smothering or choking on components; for example, whether emergency diesel generators could be adversely affected by the atmospheric dispersion of a particular chemical;
(iv) Lethal or disabling effect — personnel: The potential effects on facility personnel;
(v) Duration: The time frame during which the hazardous material is present, with an indication of whether or not dispersion occurs;
(vi) Extent of penetration: This column describes the extent to which the hazardous material migrates into buildings through flow paths, including heating, ventilation and air-conditioning systems, or whether the hazardous material remains in the facility yard.

III–10. Figure III–6 provides an example of an extreme environmental load definition matrix, containing the load environments and load combinations for engineering evaluations. The columns in Fig. III–6 are labelled and defined according to the above tables.

REFERENCE TO ANNEX III

[III–1] INTERNATIONAL ATOMIC ENERGY AGENCY, Seismic Hazards in Site Evaluation for Nuclear Installations, IAEA Safety Standards Series No. SSG-9 (Rev. 1), IAEA, Vienna (2022).

IAEA
International Atomic Energy Agency

No. 27

ORDERING LOCALLY

IAEA priced publications may be purchased from our lead distributor or from major local booksellers.

Orders for unpriced publications should be made directly to the IAEA.

Orders for priced publications

Please contact your preferred local supplier, or our lead distributor:

Eurospan
1 Bedford Row
London WC1R 4BU
United Kingdom

Trade orders and enquiries:
Tel: +44 (0)1235 465576
Email: trade.orders@marston.co.uk

Individual orders:
Tel: +44 (0)1235 465577
Email: direct.orders@marston.co.uk
www.eurospanbookstore.com/iaea

For further information:
Tel. +44 (0) 207 240 0856
Email: info@eurospan.co.uk
www.eurospan.co.uk

Orders for both priced and unpriced publications may be addressed directly to

Publishing Section
International Atomic Energy Agency
Vienna International Centre
PO Box 100
1400 Vienna, Austria
Telephone: +43 1 2600 22529 or 22530
Email: sales.publications@iaea.org
www.iaea.org/publications

24-01860E-T

FSC
www.fsc.org
MIX
Papier | Fördert gute Waldnutzung
FSC® C083411

Kolibri 360 GmbH
Ferdinand-Jühlke-Straße 7
99095 Erfurt, Deutschland
produktsicherheit@kolibri360.de